The One-Day Expert Series

IMPLEMENTING STANDARDIZED WORK

Measuring Operators' Performance

The One-Day Expert Series

Series Editor

Alain Patchong

Implementing Standardized Work: Measuring Operators' Performance

Alain Patchong

The One-Day Expert Series

IMPLEMENTING STANDARDIZED WORK

Measuring Operators' Performance

Alain Patchong

CRC Press is an imprint of the
Taylor & Francis Group, an **informa** business

CRC Press
Taylor & Francis Group
6000 Broken Sound Parkway NW, Suite 300
Boca Raton, FL 33487-2742

Printed in the United States of America on acid-free paper
Version Date: 20120703

International Standard Book Number: 978-1-4665-6328-5 (Paperback)

Library of Congress Cataloging-in-Publication Data

Patchong, Alain.
Implementing standardized work : measuring operators' performance / Alain Patchong.
p. cm. -- (The one day expert)
Includes bibliographical references and index.
ISBN 978-1-4665-6328-5
1. Standardization. I. Title.

HD62.P38 2013
658.5'62--dc23 2012024326

Visit the Taylor & Francis Web site at
http://www.taylorandfrancis.com

and the CRC Press Web site at
http://www.crcpress.com

Contents

Acknowledgment

The One-Day Expert series is the direct consequence of my previous work at Goodyear. I owe thanks to several of my former colleagues who provided me with valuable remarks and comments.

I am thankful to Mike Kipe who was part of the team I formed to deploy "Standardized Work" in Goodyear plants. He also brought some ideas to the table he got while interacting with Sumitomo Rubber Industries, Ltd. I therefore extend my thanks as well to Sumitomo Rubber Industries associates for sharing their knowledge with Goodyear.

Throughout the writing process I received numerous ideas and suggestions from Dariusz Przybyslawski. Besides being part of the deployment team, his help was instrumental in structuring and tuning The One-Day Expert series as well as for establishing the proof for concept of such a project.

I am very grateful to two other former colleagues at Goodyear who, at a very early stage, believed in Standardized Work as presented in this book, and gave me the opportunity to try it on the shopfloor: François Delé and Markus Wachter.

I am obliged to Xavier Oliveira, Pierre-Antoine Rappenne and Philip Robinson, who have read drafts and offered valuable suggestions for improvement.

I would also like to recognize and thank some of my past and current bosses for letting me explore the improbable challenge of working for a company in an operational role while teaching and doing some research: Hugues Hebbelinck, Serges Ripailles, Christophe de Baynast (at PSA Peugeot Citroën), Jacques Parfait, Dan Ludwig, Serge Lussier (at Goodyear), Christophe Aufrère, Hagen Wiesner and Patrick Koller (at Faurecia).

I express my gratitude to all my colleagues at Faurecia who work with me in testing ideas and actions, and critique or support my thoughts.

Finally, and most especially, I would like to give my special thanks to my wife, Patricia, and my son, Elykia, for their unrelenting support and patience.

Preface

THE ONE-DAY EXPERT

The One-Day Expert series presents subjects in the simplest way, while maintaining the substance of the matter. This series allows anyone to acquire quick expertise in a subject in less than a day. That means reading the book, understanding the practical description given in the book, and applying it right away, in only one day. To focus on the quintessential knowledge, each The One-Day Expert book addresses only one topic and presents it through a streamlined, simple, narrative story. Clear and simple examples are used throughout each book to ease understanding and, thereafter, application of the subject.

1

Introduction

Operators are the ones who create value in a manufacturing company. Their performance is instrumental in the success of such organizations, particularly in high labor cost countries. Measuring the performance of operators is a central piece in their improvement. As Lord Kelvin once said, "You do not improve what you cannot measure." When people think about measuring the performance of operators, they generally cling to a stopwatch and tend to focus on the quantitative aspect of the operators' work. Not surprisingly, it is usually admitted that the operator who works the fastest is, de facto, the best performer. But, let's not forget the often-overlooked basics: Operators are not robots.

This book, the first in The One Day Expert series dedicated to Standardized Work, is about operator performance measurement. In this book, we explain how to measure the performance of operators quickly and simply, how to identify the most efficient operators, and how to monitor their improvement over time. The method described can be automated very easily, thereby requiring no labor consumption. The approach is grounded in one of the basic laws of factory physics: mastery of variability. The method presented here has been applied with success for years in the automotive industry.* However, it should be underscored that this method is only a tool that should be used in the broader framework of Standardized Work deployment. We believe that Standardized Work, which is done *with* operators *for* operators, is a powerful vehicle that contributes to fostering and disseminating knowledge. Standardized Work deployment is the DNA coding of operational excellence—process by process, workstation by workstation, sequence by sequence, wherever there is a human being. It is all about finding and applying the best operational method that will

* The author conducted the deployment of an early version of the method in PSA Peugeot Citroën plants and a more complete version in Goodyear plants.

lead to cost reduction, better product quality, and increased operator safety. This contributes to progress company-wide through augmentation of knowledge and betterment of operator work conditions and morale. Ultimately, because the "right process leads to good results," Standardized Work success will have a visible impact on the bottom line of the company's financials. The real beauty of the entire story is that Standardized Work requires virtually no investment. Hence, it really epitomizes the core definition of Lean: "Doing more with less."

In this episode of The One-Day Expert, Thomas, a plant director in an industrial group, is reassigned to another plant, which is losing money. Previous plant managers have tried several initiatives with, to say the least, limited results. His urgent mission is to turn the plant around. The morale in the plant is very low and the staff is equally pessimistic about the plant's future and is distrustful of senior management. Time is running out; company headquarters needs concrete results and has become impatient. To face these challenges, Thomas has decided to use Standardized Work deployment to achieve quick and visible results while rebuilding a real team. This book recounts these initial steps of the Standardized Work deployment. Additional steps will be detailed in proceeding books in this series.

Standardized Work deployment is the DNA coding of operational excellence.

2

A Strange Day in a New Plant: The "Beheaded Chicken"

Landing at Charles De Gaulle airport in Paris was delayed due to an early-morning strike of air traffic controllers. "That's France" came immediately to Thomas's mind. Despite a very promising forecast, the weather was dismal at the airport. Late winter had brought a lot of wind, clouds, and rain this year. All of this impacted Thomas's thoughts about his assignment. Many questions came into his mind: "Why isn't the plant performing? Are unions very strong? Are people involved? Do they want to change? Will they accept me? And, finally, "Will I succeed?" Another reason for Thomas's anxiety was that the previous plant director had been fired five days ago. Therefore, he would not receive induction from his predecessor.

On the first morning in the plant, Thomas decided to take a quick tour of the shop floor. A home appliance plant, three separate departments produce the goods made here. A manager heads each department and reports to the plant director. The three departments are preparation, assembly, and final finishing. The preparation department includes four main activities, grouped in four areas: frames (frames welding), electronics (electronic components preparation), body (metal sheet stamping), and plastic (plastic molding). This last department was organized as a job shop and the production of each reference was done in big batches due to lengthy change times. Some of the changes here would require a total cleaning of the machines, such as for plastic molding. This would result in one to two hours of stoppage.

The assembly department had three areas. In the first stage, frames and electronic components were assembled. This stage had been identified as the bottleneck of the plant. Several operators in parallel workstations, as in the other assembly stages, performed operations here. In the second stage, a subassembly coming from the first stage was assembled with steel sheet. In the third and final assembly stage, plastic parts produced in the preparation department were added to the subassembly coming from the second stage. After leaving the assembly department, the home appliance was sent to the finishing department for final inspection and packing.

Back in his office, Thomas did a very quick sketch of the material flow (Figure 2.1). He noticed that it corresponded exactly to the synthetic flow chart he had received from Steve, the plant's industrial engineering manager, a few days earlier. "This is a good point to start with," he whispered with some sense of relief.

The plant had a few fundamental recurrent issues: production was not done according to the schedule, never on time, at too high a cost, and not always with the best quality results. Waste in this plant was the highest in the entire corporation. These issues had huge consequences on the plant financials because they translated into a substantial loss of materials, and the cost of raw plastic had been going up consistently as a consequence of increased oil prices. "I already turned around factories that had some of these problems, but none with all of them at once. This new place seems to be a real challenge," Thomas thought.

His observations on the shop floor were quite typical of an organization facing such problems: high inventory in many places and shortage of semiproducts in other process steps, as well as limited, not existing, or poor instructions at the workstations. Operators were performing their jobs in their own way. All of this resulted in many process flow disturbances.

Work in this factory was labor-intensive despite some automation of the process. Also, labor costs contributed a lot to the overall cost structure. Therefore, Thomas decided he would focus on people. "Everything is in their hands … and probably in their minds too," he thought.

Thomas's next day was dedicated to the review of operators' work. He was appalled by what he saw during his visit—operators doing a lot of motion. Was that really an efficient way to do things? He doubted it. Probably some of them wanted to demonstrate how much effort was required on their job. In most cases, that was true. Components in the assembly department were located far from the machines, and each operator had to walk quite a lot during the shift. In addition, the same workstations with the same machines, delivering the same product, were arranged in different layouts. That was another factor influencing unequal outputs.

Thomas was surprised that such basic Lean principles had not reached this plant. He remembered that the Lean/Six Sigma corporate director

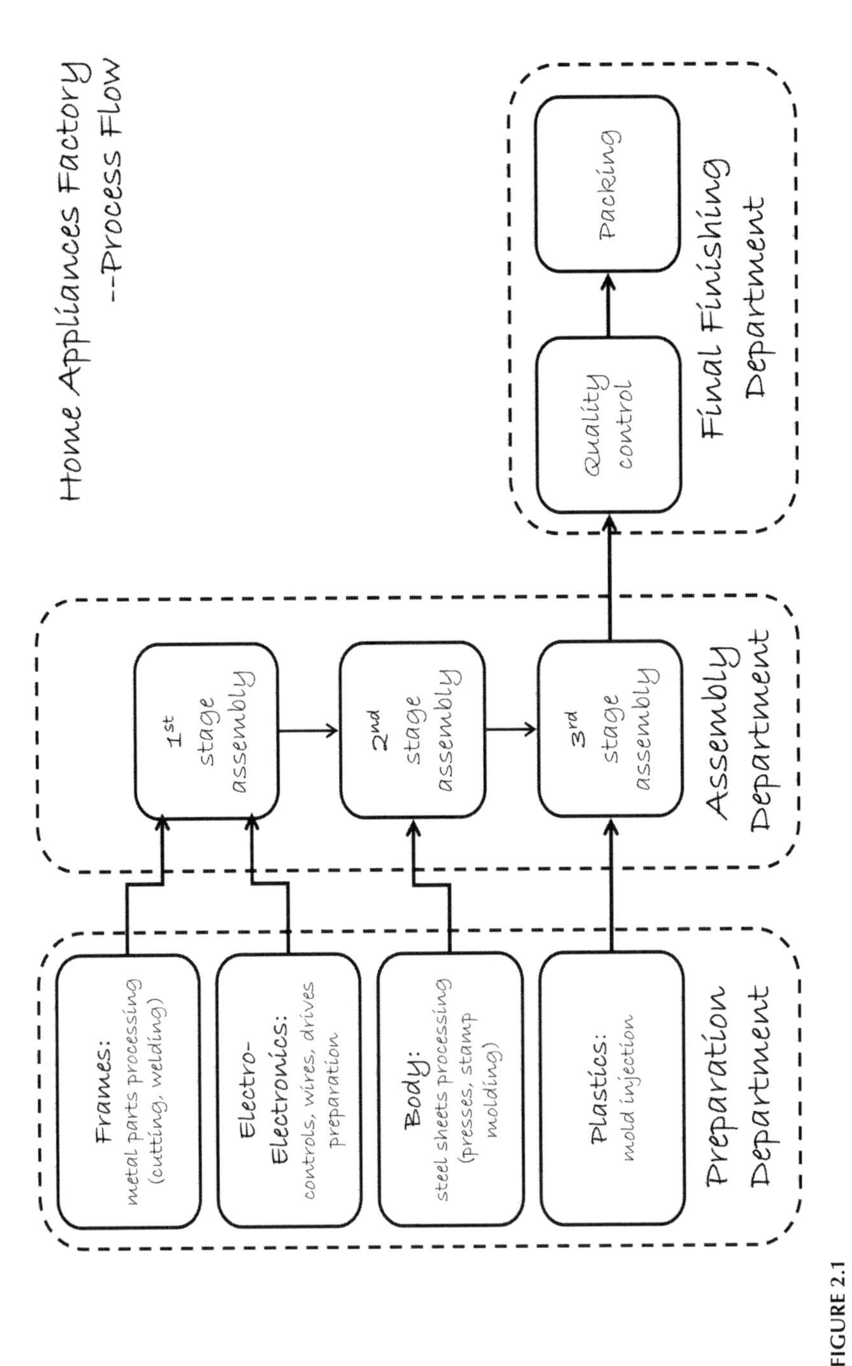

FIGURE 2.1
Material flow sketch for Home Appliance Factory.

gave a very nice presentation at February's directors' meeting to explain the huge progress they had achieved in the plants. "Just numbers." … This reminded him of the famous saying: "When you manage by numbers, they always improve." The director also talked a lot about how the massive new Six Sigma training he was launching would transform plants worldwide. The training he referred to actually consisted of giving a group of selected people several hundred hours of training on statistical tools. Thomas had serious doubts about the efficiency of such an approach. As he stated it, "Those statistical tools are so sophisticated that most of those guys would end up using less than 5% of their learning and time on real problems. After all, Six Sigma seemed to be the flavor of the season and, therefore, had to be spread.

For sure, he would willingly concede that experts could definitely use the Six Sigma tools to solve some complicated problems. However, he strongly believed that there were a lot of problems that could be solved using very simple tools. Therefore, people should be focusing on the simple things first, he thought.

Due to numerous meetings with his new team, the end of the week was busy for Thomas, but two fundamental questions didn't give him any peace: What to do? and Where to start? Unfortunately he couldn't come up with an easy plan. Everything was too scattered for him to be able to focus on any one thing. In addition, the entire next week was dedicated to various meetings regarding headcount. "What a waste of time. Again," he thought. The headcount here was a very big deal. Because the productivity of the plant was among the worst in the region, the corporate controllers decided that they would focus like a laser on headcount. They wanted the plant to cut, cut, cut. It looked like cutting without changing the way of doing business was simply making things worse over time. This situation sounded so irrational. Because of the delicate situation of the plant, regional management made it a "focus plant." This meant more and closer reporting, which created more demand for information. Because the plant did not look great, local plant workers needed to spend more time to prepare reports and make them more presentable. The plant management was assailed by checkers' and cross-checkers' e-mails, more and more questions every day. More prosaically, taking care of reporting, answering e-mail questions requested to organize meetings, meetings to collect data, meetings to have them validated by the midlevel management, meetings to get the plant manager approval for external communication, and on and on.

Meetings multiplied because the plant manager had instructed his entire team not to communicate any data out of the plant without his approval.

Above all, they were receiving lots of visits from the region. Everyone has a good reason to visit. Technical people travelled to the plant with the intention of support. Regional senior management wanted to convey their messages and show that they were active in solving problems. Preparing for visits and making the plant look good took a lot of time. All of these new tasks kept the management away from the shop floor where they were needed to solve the problems of such an ailing plant. In the end, as one engineer put it: "The more the regional team 'supported' us, the worse things seemed to get; we worked more and more every day, but we could not see any improvement, the situation was simply worsening." The plant had entered a death spiral. It looked like "the medicine was killing the patient."

Under very high pressure from headquarters, the EMEA* regional operations vice president decided to move his best "general" to the plant. Thomas was a young and very promising plant director whose present factory was performing very well. He was viewed as having some of the highest potential in the EMEA region. This was probably why the regional operations vice president chose to reassign him to this very difficult plant. His present plant was smaller and he had managed to get the best guys around him to run the factory in a smooth and stable way. All of the sudden, he would soon have to meet different challenges of a different magnitude. "For sure, this is a defining step in my career; the time has come to show that I have the guts to make tough things happen," he kept saying.

Thomas was the fifth director in less than six years in this plant, which also had gone through a lot of management changes. However, all of those changes were not establishing any integral stability. Extensive middle management reorganization had entirely scattered many people's positions as well. As is often the case, the regional senior management had started to play the finger-pointing game early on when the first problems emerged. No need to say that middle management and shop floor supervisors were somewhat disoriented and not sure of the direction the plant was heading. There also had been some doubts recently concerning the future of the plant. Would headquarters be patient any longer?

---

* Europe Middle East and Africa.

It looked like no one could find the key to unlock the potential of this plant. The region was definitely trying very hard. People in the plant were equally trying very hard. The level of mess on the shop floor showed that everything was going every which way. The image that best represented the situation in Thomas's imagination came from his childhood as a son of a farmer: a beheaded chicken. "Yes. This plant is really running like a beheaded chicken," he thought.

3

Variability Is the Enemy of Production

Spring was slowly coming to the little town where Thomas's plant was located. This was a nice and friendly place to live, but Thomas hadn't had the time to enjoy the beauty of the nature in the surrounding area. His time was mostly split into hours spent in the plant and short nights in his hotel room. His wife and their two sons were still in Austria. Because of the workload, he had not been able to find a decent place for his family to move into when they came. His weeks were now very typical: arrival on Monday morning, spending most of the time during the week in the plant, and departure on Friday afternoon. Staying in a hotel away from his family gave him the opportunity to focus almost full time on his tough new assignment. This was a bittersweet situation because the time dedicated to the plant came at the expense of his personal life. Forced by the situation, he was reduced to a maxim so often heard from a long-time colleague who talked about the time spent with his family: "If you cannot get the quantity, increase the quality." In application of that maxim, Thomas decided to be fully dedicated to his family on the weekends and to have more leisure time with his two young sons.

He had been around for weeks now and, despite being almost 100% dedicated to the plant, he had not yet found any good solutions. Fortunately, a few days off helped to put his first thoughts together and make some analysis. Bottlenecks were inconsistent and sporadic for different reasons. Two of the influential factors were unsteady operators' performance and equipment breakdowns. It was obvious that this situation created a lot of variability in the process, which disrupted the flow of materials. After returning to the plant, he collected some data and made a quick study, which confirmed that machines were partially at fault, but their impact was not as important as the operators' impact.

Two weeks later, Thomas understood that he needed to reduce variability in his plant, especially in the bottleneck areas. He also came to the conclusion that he needed a plan to inspire and build teamwork in the plant. The structural problems were just one side of the coin; it would take the right approach and the right organization to move forward. It became obvious for Thomas that he needed the people to be behind any initiative, and needed to involve them. He always remembered what talented leader Sam Walton said: "Outstanding leaders go out of their way to boost the self esteem of their personnel. If people believe in themselves, it's amazing what they can accomplish." In this plant, the people's self-esteem was full of wounds that he needed to heal. He also wanted all of the operators to be part of the process. "They should understand that we are doing this to support them, not to squeeze them and get the most out of them. They should be convinced that we are all in the same boat. If they are not onboard, we will fail." How could he begin?

At this time, he thought about a guy in the region who had been advocating the deployment of a project called *Standardized Work*. "What was his name again?" Daniel, Daniel something … Daniel Smith. He had been recently hired by the company to be in charge of Industrial Engineering in the region. He didn't have the time to learn a lot about all of the plants in the region, therefore, his knowledge of the process was somewhat limited, but Thomas didn't think this was really a big issue. The centerpiece of his initiative was about reducing variability. He had heard him several times saying that "variability is the enemy of production." The high level of variability was definitely a problem in his new plant. He decided that he would contact Daniel to discuss how he could help support his plant. Thomas was a "do-it-now" person who doesn't understand procrastination. He, therefore, decided to call Daniel right away. The phone call was very quick. He went straight to the point, and explained the situation in his plant and his intentions. Daniel, who was looking for enthusiastic plant managers to show the power of his Standardized Work initiative, was not difficult to convince. They agreed to organize a training session on the subject in his plant. Thomas wanted a massive impact on the production and it was important for him to make sure everyone who was going to be involved also would be trained. Therefore, he decided to pull from the daily business 32 of his best people for the training, which was scheduled for eight days later.

Daniel explained his "one-day expert" approach. It consisted in giving simple but sufficient knowledge to people in one day so they could be able to go on the shop floor and start doing things right away with minimum supervision. He had a catch sentence for that: "Everyone, Everyday, Everywhere." He called that the 3E. Thomas was really seduced by this approach, which contrasted with the whole Six Sigma philosophy. "This is another common ground between us," he said.

Another thing he liked in Daniel's approach was that it led to very low CAPEX* actions. There also was something even more important for Thomas—3E was centered on people and enhanced teamwork. This surely was not another tool-of-the-season, but a system built around people. Thomas was convinced that this would help create successful organization to support the plant's turnaround. Daniel explained to Thomas that to deploy Standardized Work, he had built a regional support team including the regional training manager as well as a representative of regional Excellence System† team. "Teamwork again in action," exclaimed Thomas.

> *Variability is the enemy of production. Operational excellence is all about identifying then eliminating or absorbing variability the smartest way.*

---

* CAPital EXpenditures.

† Excellence System was the company's Lean deployment initiative.

4

Training Day One: Introducing Standardized Work

June 4th was a very sunny day in this small city near Paris. Daniel drove from his place close to the regional headquarters' office the day before to make sure that the training preparation would start at 7 a.m. the next morning. He had a lot to do before the start of the training at 9 a.m. The trip took roughly four hours. Therefore, he had plenty of time to think about the training and some key messages he needed to convey. He had led several training sessions in the company's plant, but this one was really special in terms of expectations and attendance. As always, the message he wanted to underscore was that he was simply there to support the plant and that the people who would really be making changes were the trainees and the rest of their colleagues.

This Monday was the first day of training for one of the four eight-person groups Thomas had formed. Daniel had had several phone discussions with Thomas to set up training sessions and make sure that each of these groups was made up of people with different backgrounds: production, engineering, maintenance, safety and ergonomics, industrial engineering, and the Excellence System.* They also picked a leader for each group. Before finalizing the groups, they also checked that members were basically compatible.

Daniel arrived early in the morning in the plant to start the preparation. He had some standard approaches of his own on preparing training rooms. First, he would post a few charts with key messages he would share with the trainees. Second, he will post squared, blank poster papers, which would be used by each group to draw charts while answering questions during the training. The final step was always dedicated to setting up the stage for the simulation. Because he always wanted things to go smoothly, as if scripted, he had to pay attention to every detail. His two mantras on the

* Excellence System was the local version of Lean deployment.

training were: (1) "leave no place to aleatory (random) outcome" and (2) "the right process will lead to the good results."* After slightly less than two hours of preparation, the room was almost ready to begin the training.

Thomas arrived in the training room 15 minutes before the start. He wanted to personally kick off the event, thereby showing all participants how important this was to him as well as to the future of the plant. Daniel had just finished sticking the latest papers on the walls when Thomas appeared in the room. The training started at 9 a.m. sharp as planned. Thomas took the stage first. He spoke for about 15 minutes. After briefly introducing Daniel, he went on to explain the reason for his presence. He presented the training as the starting point of the deployment of the Standardized Work. He then explained the importance of Standardized Work in all improvement initiatives. "We are very lucky to have Daniel here to support our plant, so please pay attention, make sure that you understand, and, if not, ask questions." Thomas then turned it over to Daniel to begin the training.

After a quick roundtable of succinct self-introductions, Daniel started by discussing the question most of the trainees had on their mind despite Thomas's kickoff: "Why are we here? Why is Standardized Work so important for us in our daily job?" As always, Daniel began by asking a few questions whose answers would help him illustrate "Why Standardized Work?"

The first story was about improvement made on an assembly machine. He always took an example of a machine from the host plant. "As you know," he said, "Excellence System people are currently very busy helping plants in the region reduce tool change time. This is what we call Quick Changeover or QCO. Let us suppose that you have got an assembly machine where you would like to implement QCO. What would this machine be?" Daniel moved to a nearby flowchart and took a marker ready to take notes and draw (Figure 4.1). After a few moments, Steve, the Industrial Engineering (IE) manager, responded:

It would be the M1.

Daniel: What is the daily production of this machine?

Steve: Well, this machine produces roughly one part every minute. That makes a daily production of 1152 parts.

Daniel: Okay, what is the total change time per day on this machine?

Steve: Umm … I would say two hours.

---

* This was Daniel's formulation of the famous Toyota saying: "The Right Process Will Produce the Right Results."

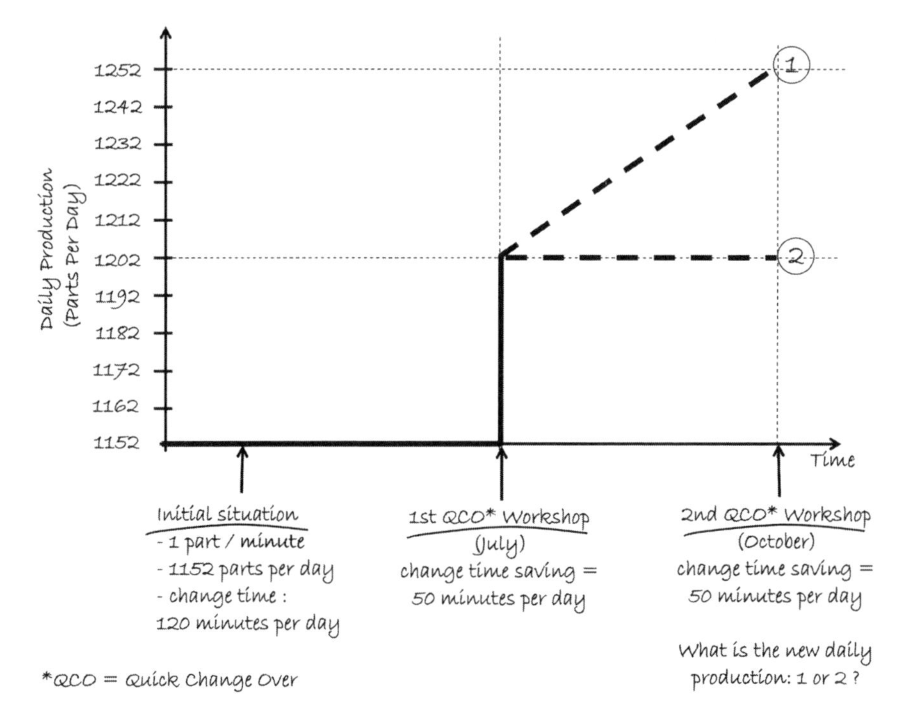

FIGURE 4.1
Changes in daily production resulting from QCO workshops.

Daniel: Alright, suppose that you decide to implement a quick changeover on this machine. You then call an Excellence System person from the region to help you. The first workshop is held in July. During the workshop, you manage to develop a new change method that reduces the daily change time by 50 minutes. What will be the new daily production of the machine at the end of this first workshop?

Steve: Well, since the machine is producing a part per minute, the new daily production should be 1152 parts plus 50 parts. That makes 1202 parts per day. Right?

Daniel: Good. Now let us suppose that the regional person who supported you for the workshop got another nice idea for the QCO and proposed to organize a second workshop. This workshop is performed in October. The result of this second QCO workshop is another savings of 50 minutes per day. What will be your daily production after this workshop? Will it be 1202 parts per day or 1252 parts per day?

Steve: Umm …

Daniel: Do you have any idea?

Steve: I would say 1252 parts per day. It seems so obvious. But, the fact that you are asking the question tells me that it is not the correct answer.
Daniel: Well, you are right to be quizzical. Who else can answer this question?

The room stayed silent for a few seconds. Daniel broke the silence with. "The truth of the matter is that we do not know. The answer depends on what you will have done between the workshops. With no Standardized Work implementation, I mean such as writing down the new method, training people, and auditing, you can be sure that the production will be around the initial number: 1202 parts per day. What is the bottom line here? Well, as Taiichi Ohno* once said, 'With no standard there is no improvement.' Let me underscore this again: Improvement alone is a waste of time if there is no standard. Now, don't get me wrong. I am not proclaiming the supremacy of the standard over improvement. For sure, standards prevent improvement from evaporating, but, on the other side, the raison d'être of a standard is to nurture and record improvement steps. At the end of the day, standard and improvement are the two sides of the same coin. They need each other to last."

Daniel recounted two other stories to illustrate the need for Standardized Work. He wanted to conclude this segment with a reminder. Therefore, he went to the flipchart and requested the room's contribution. "Could you please help me summarize why we need Standardized Work?" He wrote down the answers and then summarized. "According to your answers, Standardized Work will, first, help sustain improvements across all shifts; second, it brings the stability needed to implement other Lean and Excellence System tools; third, it is a medium to harvest and spread out best practices; fourth, it allows people to see normal versus abnormal conditions, solve problems, and improve." He found this feedback quite in line with his teaching. Hence, he decided not to add any additional points and congratulated the group for their quick understanding (Figure 4.2).

Now that he had made the case for Standardized Work, Daniel thought the time was right to introduce what he called the *training package* (Figure 4.3). He explained the Standardized Work training had five phases and gave the rationale and the details regarding each one.

"Phase one consists of classroom training. This is about concepts and tools presentation. Phase two is a simulation module whose goal is to materialize the previous phase. Please note that to make training more efficient,

* A Toyota engineer who is considered to be the father of the Toyota Production System or Lean Manufacturing.

Standardized Work

- Sustains improvements across all shifts
- Brings the stability needed to implement other lean and Excellence System tools
- Is a medium to harvest and spread out best practices
- Allows people to see normal vs. abnormal conditions, solve problems and improve

FIGURE 4.2
Main benefits of Standardized Work deployment.

the delivery of Phases one and two will not be sequential, but intertwined. Phase three is the on-the-floor practice. The goal here is to finalize the materialization and bring to life the learning. The trainee will have the opportunity to apply the tool on a process he or she knows very well. He or she will meet unexpected and concrete issues that will need to be resolved. After this phase, understanding and confidence grow. This phase epitomizes the "learning by doing" approach. Phase four is implementation on the shop floor. In this phase, the plant is expected to replicate the learning on the shop floor along with some timely coaching from the trainer. That means from me or from my colleagues from the region. Here, the trainee will face additional challenges. Success will require both motivation and mastery of tools. Now, it should be clear that the ultimate justification of any concept deployment is the link between its success and the key performance or business indicators. Phase 5 seeks to show and validate the linkage."

Daniel insisted that, before any deployment, this link needs to be shown, or agreed upon, and later validated. Now very concretely, you should expect the full deployment of Standardized Work to increase your productivity, reduce your cost, and increase the quality. What's more, you should notice an increase in safety and the morale of people in your plant. At the end of the day, you should be able to check and validate the link between Standardized Work deployment and the betterment of each of the previous indictors. He then concluded, "There would be no sustainable deployment without this link." Daniel also explained that the week's training would only cover Phases one to three. As for Phases four and five, he underscored that "the plant will be in charge." He and his regional colleagues would only be doing some coaching at designated and agred intervals. "Why do we really need to go through all of those phases?" Daniel asked the audience. One of the participants, a young lady who was from the training department responded that each phase would increase the level of retention: According to her, the level of retention generally grows from around 5% when a trainee attends a lecture to around 90% when this lecture is followed by an application of learning in a real situation. The training coordinator then concluded, "Ultimately, the best way to learn is by doing, everything before helps you get there." Daniel confirmed the answer and completed "only seeing a few slides on Standardized Work will never make you an expert. To better master the subject, you need to go through multiple steps. The training will be considered good enough only when you are able to do it alone on a real example. Even then, you are not 100% trained. The rest of the learning will come with the experience you will acquire treating many cases."

> *The ultimate justification of any concept deployment is the link between its success and the key performance or business indicators. Before any deployment this link needs to be shown, or agreed on and later validated. No sustainable deployment is possible without this link.*

After this brief introduction to the training package, Daniel gave an overview of the training week (Figure 4.4). "I do not want to load your mind with lots of details at this stage. Therefore, for the time being, I will only give you the main steps of the Standardized Work. Those steps are pretty straightforward. We will start by training on a few tools that you will need to capture the current state. These tools are: Operators' Performance Mapping and Standardized Work forms. At the end of this introduction we will see the first of those tools, which is on Operators' Performance Measurement. After this initial step on 'capturing the current state,' we will see how Standardized Work can be used to foster improvement with a special focus on some quick and simple methods to improve. Then, we will follow the 'training' step. We will see that it is a key point in implementing Standardized Work successfully. 'Auditing' is the practice that will help you check if your deployment is actually a success or a failure. More importantly, auditing will help you continuously improve your Standardized Work deployment. Please remember this is not a 'finger pointing,' but a 'problem-solving' approach. We do not perform auditing to punish operators who are not following the defined Standardized Work. As we will see later, the objective is clearly to find the 'why' and solve the problem. Now, in terms of planning, we will start with the Operator Performance Mapping today. We will continue the training on this tool tomorrow. Wednesday will

Standardized Work Package

Phase	Content	Delivery method	Confirming learning
I and II intertwined	Classroom STW* training	PowerPoint, graphs, drawing explanations	Quizzes
	Simulation	Model where a factory is packing T-shirts. The group works trough all the steps of STW explained previously.	Observation and feedback by facilitators
III	On the floor practice	The group works trough all the steps of STW on selected actual case on the shop floor.	A-3 completed as a result of the activities, presented to managers and shared with regional team**.
IV	Implementation	Plant responsibility	At designated intervals, regional team reviews. Achievements are identified and shared with other plants. Gaps are addressed with coaching and development.
V	Linkage with key performance and business indicators	Plant responsibility	At designated intervals, with regional team impact of the STW on the key performance and business indicators is reviewed and assessed.

*STW = Standardized Work
**Regional team includes IE, Excellence System and Training.

FIGURE 4.3
The five phases of the Standardized Work training package.

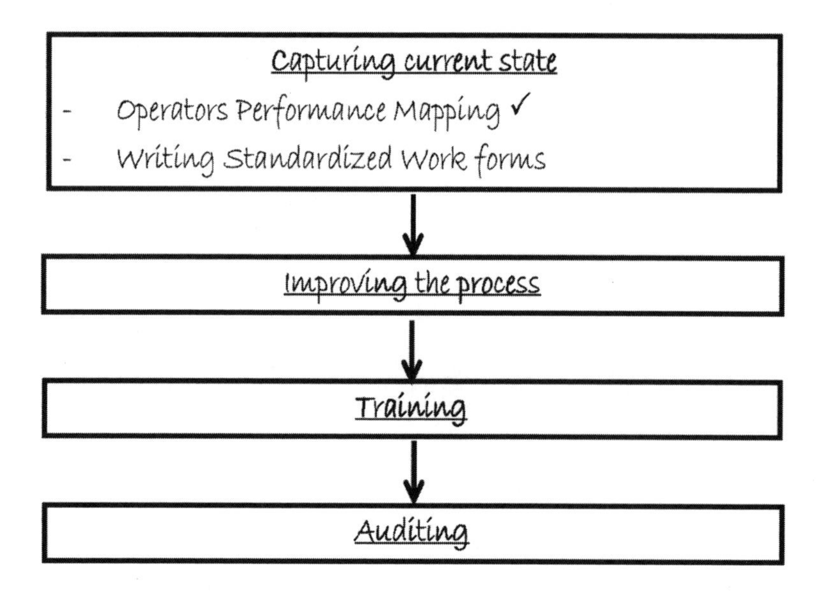

FIGURE 4.4
The four steps of Standardized Work deployment.

be devoted to the writing of Standardized Work support documents. Among them there are a Process Analysis document, a Standardized Work chart, a Standardized Work combination table, and an Operator Work instructions sheet. Then, Thursday will be dedicated to Process Improvement. We will conclude the training on Friday on two key activities needed to sustain Standardized Work: training and auditing."

You should expect the full deployment of Standardized Work to increase your productivity, reduce your cost, and increase the quality. What's more, you should notice an increase in safety and the morale of people in your plant. At the end of the day, you should be able to check and validate the link between Standardized Work deployment and the betterment of each of the previous indictors."

As Thomas was concluding this part, a participant from the training department, raised his hand and asked: "Daniel, you seem to be mixing tools that are supposed to be under the responsibility of different departments. For instance, standards should be done by IE people, process improvement by Excellence System people, training by HR people, auditing by supervisors. Please could you explain to us why you are regrouping all these tools under the umbrella of Standardized Work?"

Daniel felt that the time was right to talk about the need to deploy Standardized Work as a system consisting of several tools. He responded. "Here is the thing. We believe that to obtain the best results, Standardized Work should be deployed as a system built on several tools. Certainly each tool can be deployed alone and even by different departments as you mentioned. However, you will not get the full benefit of these tools if each of them is deployed individually. Moreover, the impact of each tool deployed alone will fade after a while. As you may see on these two drawings, this is what I call the 'fountain' approach (Figure 4.5). Now, as you can see on the second chart, if you can deploy a tool within a system, then not only will you get the most of it, but you also will obtain a sustained cumulative effect that will last. This is what we call the 'leave.' Here, the system approach provides the needed backbone to multiply and sustain each tool's impact. As I just explained a few minutes ago, our approach for Standardized Work deployment is the latter. It is built around a set of several tools. This training will consist in subtraining you to all of these tools. As I said earlier, today and tomorrow we will focus on the first tool, which is Operators' Performance Measurement. Well, let's get started."

THE SIMULATION

Daniel noticed that he had essentially been talking alone for a while. He dislikes getting into monologues during training. He got the impression that some trainees were getting drowsy. He, therefore, was relieved to move to the next module of the training. The simulation will certainly give lots of opportunities for people to speak, interact, and stretch their legs.

The simulation segment was about t-shirt folding and packaging. He asked the eight participants to form two groups. Then he pulled 80 t-shirts from a big bag. The t-shirts were packed into bundles of roughly 10, each wrapped with tape. Each group received 40 t-shirts. Along with

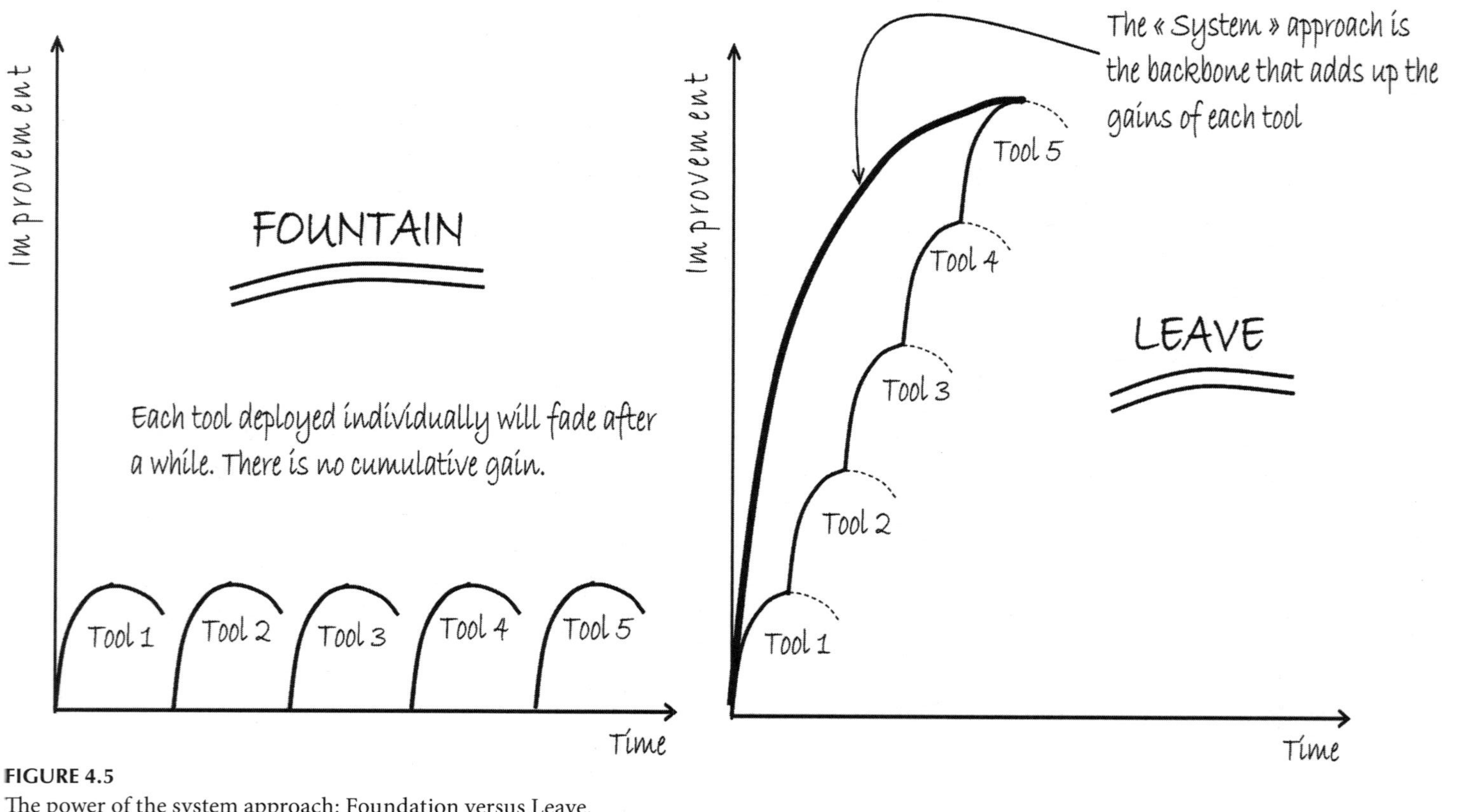

FIGURE 4.5
The power of the system approach: Foundation versus Leave.

the t-shirts, he also handed out two stopwatches to each group. Then he started pitching a fictitious story he had developed for the simulation:

> *"Our company has decided to give away a t-shirt with each appliance sold to attract more customers. The t-shirts, as you may guess, are now produced in China and we are looking for a plan to insource the production. We, therefore, have decided to send somebody to China to have a look at their current process. Here is the description of their current folding and packaging process."*

Daniel made a demo of the process, step-by-step, then continued:

> *"The goal of the simulation, at this stage, is to replicate the same process here and later we will see if we can improve it. If we are able to be more productive than the Chinese company, then the senior management would commit to insource the entire production of t-shirts. I just made a demonstration of the process as designed in China. As you can see, the process is very manual. I would ask each team to pick an operator whose role will be to fold the t-shirts as shown on the sample I placed on the table over there. I also would ask that one person on each team record the time between two folded t-shirts. I mean, every time a t-shirt is folded take note of the time displayed on your stopwatch. I usually call this 'top part' or 'TP' because it can be recorded by sending a signal every time a part is produced. If this is clear and everyone is ready, let's start. Last thing before you start, you may use the first two t-shirts to practice and if there is still something unclear for you, let me know. I will be moving from one group to another to see how things are going."*

After each group folded its 40 t-shirts, Daniel pointed to a squared poster paper on the wall and asked them to plot the time between parts. "Do not use a computer, do it manually with a marker on this paper," he insisted. "Why? It would be faster to use an Excel® spreadsheet," one of the participants pointed out. "Yes, you are right, but the point is not about going fast here. Also, it is not about me being anticomputer. This is the best way to learn and to touch the variability. Then, thereafter, you would be allowed to use the computer at your will," Daniel underscored. He continued, "A good friend of mine always told me, citing a Chinese proverb: 'Tell me and I will forget, show me and I may remember, involve me and I will understand.' There is no better way to be involved than doing it manually."

It took both groups an hour and a half to record all data and plot their graph on the paper on the wall. Daniel requested that both groups use the same paper so that the two graphs could be visualized together. This was the best way to display the differences between the two groups' results.

The last group had just finished its work when Thomas returned to the classroom to enquire about the progress of the training. It was also time for Daniel to call the participants around the two charts. He gave Thomas a succinct presentation of the work the trainees were carrying out. Daniel's teaching style was built around dialectic questioning; he very naturally started asking questions. "What do you see on the graphs, guys?"

One participant from Group 1 quickly pointed out: "We have outperformed Group 2. Our average time between parts is almost half theirs." (Figure 4.6) Daniel nodded and replied, "This is true, but what else did you notice?" Another participant jumped in and said, "There is a lot of variability in the time needed to fold a t-shirt." (Figure 4.7)

Daniel, who was very pleased with this answer, responded, "Yes!" Then continued, "Variability is your enemy. Operational excellence is all about suppressing, reducing, or absorbing variability. You must focus on that first. There are roughly two kinds of variability: special cause variation and common cause variation. Special cause variations are directly linked to special events: breakdowns, tool changes, periodical tasks, and so on. In this case, the cause and the effect are linked in a clear and obvious way. As a consequence, special cause variations are the easiest ones to investigate and to eliminate. Some special cause variations are easier to eliminate than others. These are special cause variations triggered by deterministic events like periodical stoppages or tool changes. Such events are completely predictable. Obviously from the time you start to produce in a shift, you can know or define with a good precision when you would have to change a tool, a consumable, or replenish an empty box. Therefore, no one should let himself or herself be surprised by such variations. Normally, those should be addressed in the very beginning. Gains coming from deterministic special cause variations are really like big bucks lying on the floor. You simply need to bend down and collect them. After those low-hanging fruits come the variations due to breakdowns, which are, by definition, random. Obviously, no machine is perfect and you may experience breakdown on any machine in a plant. However, TPM* and other preventive maintenance activities will substantially reduce the probability

* Total Productive Maintenance.

Time between parts for Group 1 -in seconds

Part 2	Part 3	Part 4	Part 5	Part 6	Part 7	Part 8
34	33	39	36	38	36	34
Part 9	**Part 10**	**Part 11**	**Part 12**	**Part 13**	**Part 14**	**Part 15**
36	33	34	39	36	33	43
Part 16	**Part 17**	**Part 18**	**Part 19**	**Part 20**	**Part 21**	**Part 22**
50	26	34	44	38	33	45
Part 23	**Part 24**	**Part 25**	**Part 26**	**Part 27**	**Part 28**	**Part 29**
36	36	32	40	32	46	73
Part 30	**Part 31**	**Part 32**	**Part 33**	**Part 34**	**Part 35**	**Part 36**
40	36	49	34	41	34	36
Part 37	**Part 38**					
47	34					

Time between parts for Group 2 -in seconds

Part 2	Part 3	Part 4	Part 5	Part 6	Part 7	Part 8
65	62	66	60	74	64	91
Part 9	**Part 10**	**Part 11**	**Part 12**	**Part 13**	**Part 14**	**Part 15**
58	95	90	74	63	64	67
Part 16	**Part 17**	**Part 18**	**Part 19**	**Part 20**	**Part 21**	**Part 22**
61	64	70	75	117	83	75
Part 23	**Part 24**	**Part 25**	**Part 26**	**Part 27**	**Part 28**	**Part 29**
75	73	72	119	95	68	66
Part 30	**Part 31**	**Part 32**	**Part 33**	**Part 34**	**Part 35**	**Part 36**
59	67	69	62	70	66	62
Part 37	**Part 38**					
83	74					

FIGURE 4.6
Collection of times between parts.

of failure of any machine. These kinds of actions should come immediately after addressing deterministic special cause variations. Eliminating special cause variation is instrumental in achieving plant stability, which is the foundation of operator performance.

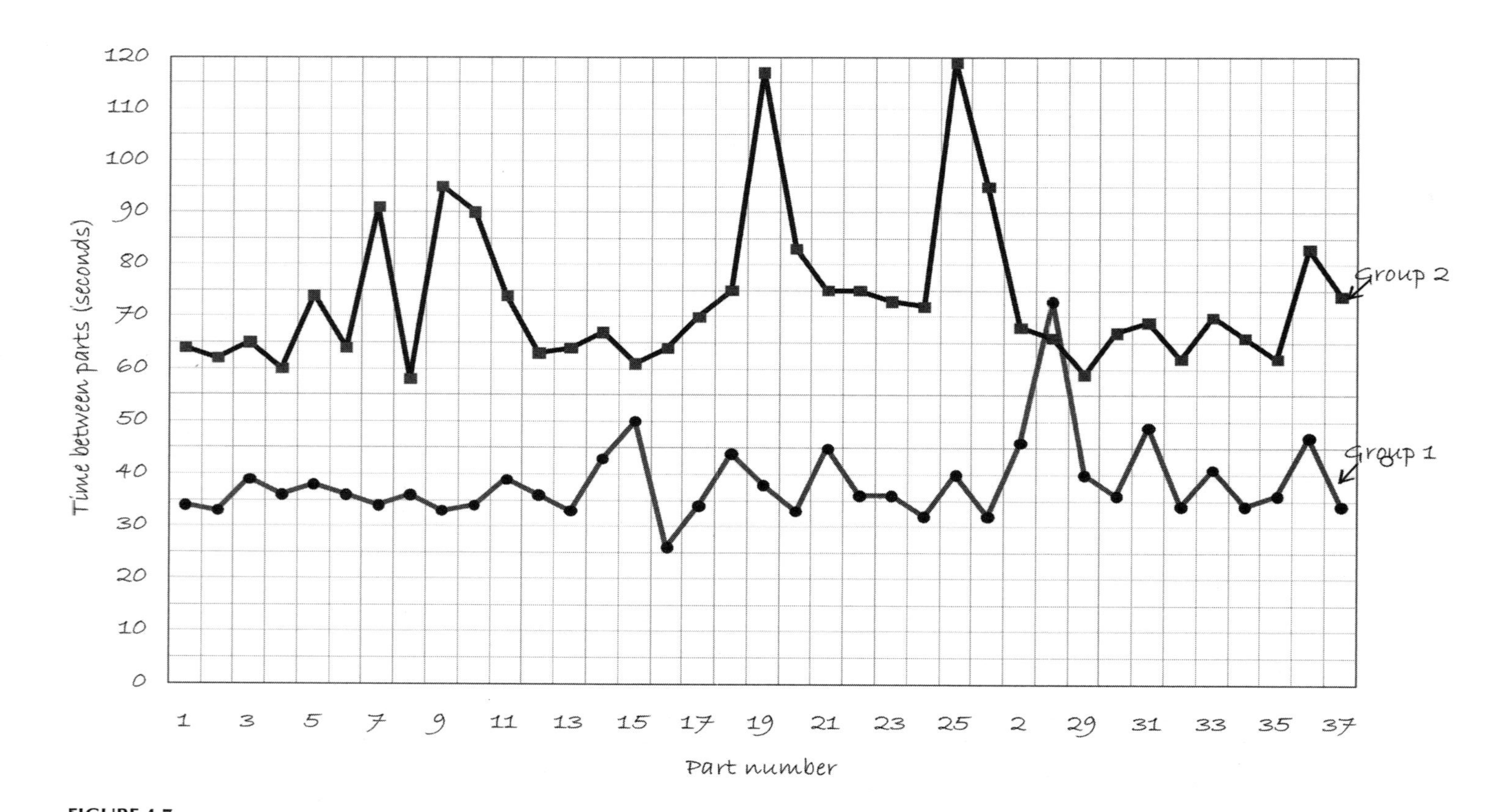

FIGURE 4.7
Visible variability on time between parts for the two groups.

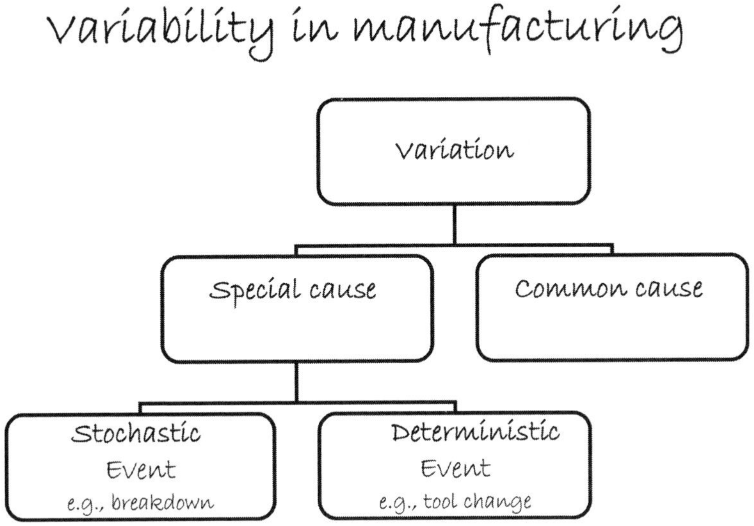

Variability in manufacturing

Variation

Special cause

Common cause

Stochastic Event e.g., breakdown

Deterministic Event e.g., tool change

FIGURE 4.8
Main kinds of variability in manufacturing.

"Common cause variations are due to a combination of fuzzy reasons, most of them naturally random phenomena. Because their causes are not easy to identify, they are also the toughest to reduce or eliminate. A classic example of such variation is the one coming from human factors. Operator performance also can be affected by special cause variation.

"Please do not try to address operator performance without achieving minimal conditions of plant stability. Doing so may lead to lots of frustration from operators. The irony is that there is a risk of not being credible in front of operators when you talk about improving their performance while they are facing huge breakdowns and other trivial stoppages. I bet that some of them will rightly point out that they are not the problem, the machines are."

Thomas, who is a very practical person, broke in. "Let me summarize in a drawing (Figure 4.8). Please stop me if there is something wrong, Daniel." He then took a marker and went to the flipchart and drew in the figure below.

> *Eliminating special cause variation is instrumental in achieving plant stability, which is the foundation of operators' performance. Do not try to address operator performance without achieving minimal conditions of plant stability. Addressing operator performance without achieving minimal conditions of plant stability may lead to frustration and demotivation of the workforce.*

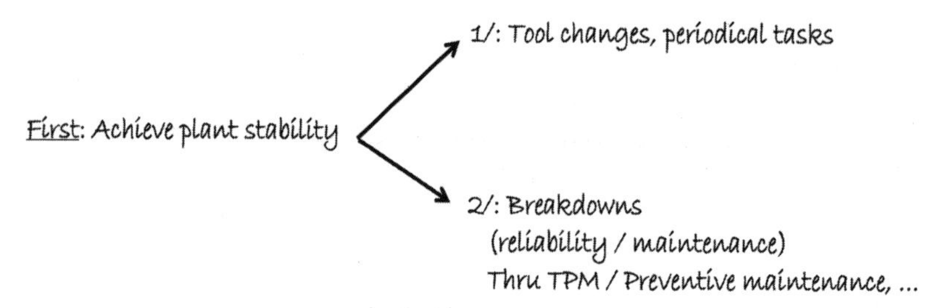

FIGURE 4.9
Priorities in addressing variability.

Thomas continued his summary with the following drawing.

All participants were looking at the figures on the flipcharts and Daniel thought that this was really a good summary of his explanations. "As always, 'a picture is worth a thousand words.' This is a perfect summary," he commented.

Thomas asked, "Do you mean that the next step would be to focus on special cause variation?" Daniel replied, "This time, we will focus on variability due to human factors." He then continued, "Indeed, as I told you previously, you should always start with the easiest one unless the process is very manual and my understanding is that you made some progress addressing plant stability so far. This leaves very little variation contribution to breakdowns and tool changes*. This is also the case here concerning t-shirt folding. That said, you should never wait until you are 100% stable before you address operators' performance. The first thing to do is to capture the current state by measuring. Then things are more like a dichotomous spiral. First, as you depicted in your chart, you should start by ensuring some process stability, this clears the floor for the initial layer of operators' performance improvement, which, in return, will make more visible some remaining process stability issues that you should fix. After that, you are back with tighter targets on operator variability reduction, and so on. Let me show it on the flipchart this time (Figure 4.10).

"Now, don't get me wrong," Daniel said, "You do not need to achieve process stability before measuring operators' performance. The point is that you should not try to improve their performance when there is no minimal process stability. Measuring operators' performance is part of

* Process analysis is the method to be used to estimate the proportion of each variation cause and is addressed in a different book of The One Day Expert series.

FIGURE 4.10
The process – operator improvement spiral.

the process of 'capturing the current state.' This is simply a common sense step that should be performed at the beginning of any initiative. This is how we would be able to measure the progress down the road."

The discussions took some time, and it was almost 11:30 a.m. when Daniel suggested that the participants take a quick break. Before the break, he wanted to make one last point, however. "In your example, human factors disruptions tend to be shorter than breakdowns. Therefore, the highest peaks tend to be special cause variations like breakdowns or periodical stoppages. The distinctive feature of the latter is that they appear cyclically. The beauty of the chart in front of your eyes (Figure 4.7) is that you can retrospectively link the peaks of those charts to special cause variation; for instance, the time you needed to remove the tape from t-shirt bundles. This is how you would start to 'see and touch' the variation in your plant."

Because you can obtain plenty of information about the process simply by analyzing the "top part," we call it the "voice of the process."

Thomas found the learning very interesting. He called his assistant and asked her to cancel all his meetings for the day so that he could stay with the group for the rest of the day. Daniel thought that it was a good thing that the plant director showed such interest, but also feared that his presence would make the other trainees quiet and subdued, and they might not be inclined to ask questions.

> *Because plenty of information about the process can be obtained by analyzing the "top part," therefore it is called the "voice of the process."*

5

The Hidden Cost of Human Variability

When the group returned from the break, Daniel took advantage of their new interest in variability and decided that he would spend some more time on the subject. He also encouraged the group to come forward with any questions without being intimidated by the plant manager's presence. After this remark, he refocused on variability: "Human operation variability operates very much like a machine breakdown. Therefore, at the end of the day, you will have to buffer it with a combination of extra capacity and extra inventory. That is a basic rule of factory physics. Therefore, it should be addressed just as anyone would in the case of frequent breakdowns, because any of these situations translate into more investment and/or more cash needed." Daniel went to the board and summed up what he had just said (Figure 5.1).

After summarizing on the board, Daniel continued. "Because people are not machines, addressing human variability is much more complex. And to top it all off, this human variability is sometimes hidden in averages and coefficients that Industrial Engineers manipulate." Daniel sketched a graph on a flipchart that he described as coming from a very interesting study he had come across.* This study analyzed the link between variability and the extra capacity necessary to buffer it. The graph† depicted extra capacity necessary as a function of inefficiency due to human task variations. He explained that the chart was built on an operator with an initial efficiency of 95%.‡ The study showed this operator would need 35% of extra cycle capacity to offset his 5% manual inefficiency. He suggested the group pay attention to the exponential shape of the curve. "The depicted chart shows that the extra capacity needed to absorb variability seems to grow

* The study refers to a work by Patchong, A., T. Lemoine, and G. Kern. 2003. Improving car body production at PSA Peugeot Citroën. *Interfaces* 33 (1): 36–49.

† This chart is based on the results of the study mentioned in footnote 1.

‡ This also means 5% inefficiency.

What is needed to absorb variability?

1/ You always need extra-capacity

2/ If not enough then you would need inventory

3/ If not enough then variability will propagate to the next stage –go to 1.

Any of these situations will translate into need for CAPEX and/or Cash

FIGURE 5.1
The cost of variability.

exponentially as we get close to 5% inefficiency absorbed. It means that any extra variability that is generated by the operator work will cost even more to absorb. This is the high hidden cost of human factors," Daniel concluded (Figure 5.2).

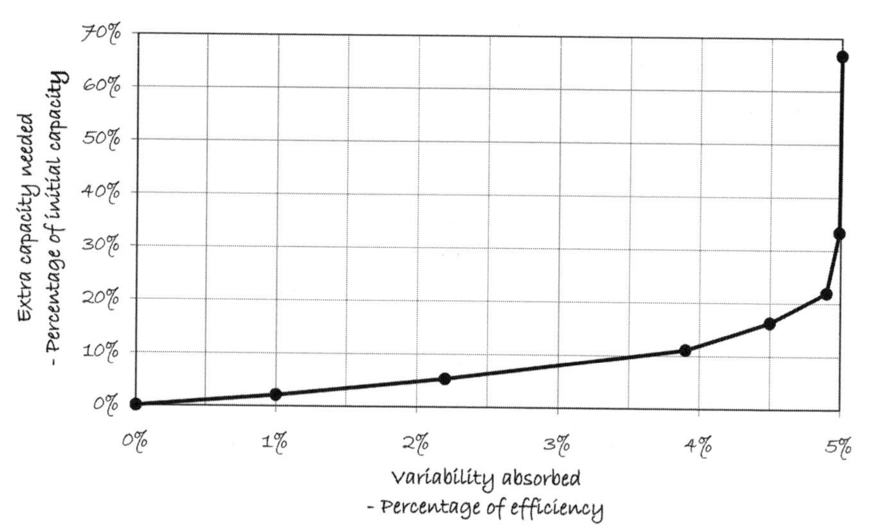

The extra capacity needed to absorb variability seems to grow exponentially: Operator initially has 95% efficiency. To absorb the whole operator variability (loss of 5%) we would need around 35% extra-capacity.

FIGURE 5.2
The hidden cost of human factors.

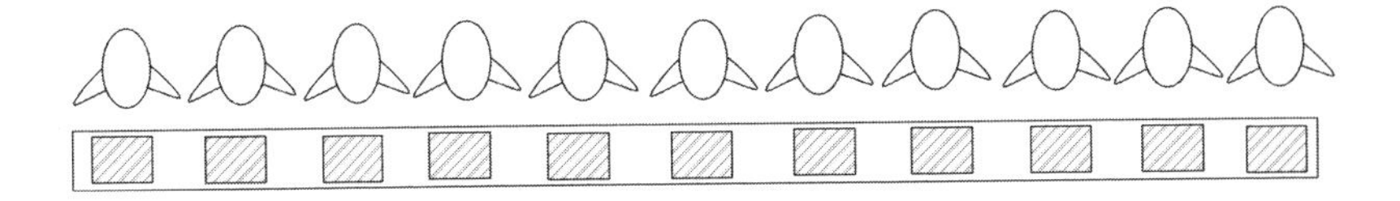

FIGURE 5.3
The cost of variability: 2% of individual inefficiency translates into 10 people hired.

Daniel took another example. "Let's suppose you have a flow line with 11 operators and that each operator is losing only 2% of efficiency to variability. Thus, 98% of efficiency seems to be a good level, right?" Most of the participants nodded. Then Daniel moved to the paperboard and wrote down the numbers. "Well, with this amount of variability, which you considered as very good, you will end up with a line whose efficiency is around 80%.[*] That means if you had no variability problems, you would have needed only nine people. In a plant like yours working 24/7, this translates into 10[†] employees being hired and paid solely to absorb variability. Now, let us be clear, you would never be able to achieve zero variability. However, keep in mind that 'variability is a killer' and you need to fight against it. Also, it has a bigger impact in Leaner plants and Leaner processes. I mean processes and plants with small inventories." The irony, as we will see later is that fighting this variability takes no CAPEX (capital expenditures). It is literally a zero-CAPEX activity.

> *Human operation variability is very similar to a machine breakdown. Therefore, at the end of the day, you will have to buffer it with a combination of extra capacity and some extra inventory. That is a basic rule of factory physics.*

---

[*] The exact formula is not $(0.98)^{11}$ but $1/(1+11 \times (1/0.98 - 1)) = 81.6\%$. However, the approximation used is accurate enough while keeping things simple. For detailed demonstration of the exact formula, visit: www.TheOneDayExpert.com.

[†] This estimate anticipates the calculation depicted in Figure 9.1, which shows that a plant that operates 24/7 will roughly need five people to fill one job position.

6

How to Measure Operator Performance

After the sequence on "the hidden cost of human variability," Daniel wanted to underscore the correlation between operator variability and performance. "There is something else very important to say about variability. When it comes to operator performance, the best performer tends to be the one with the least variability and vice versa. Moreover, as in the case of machines, the worker with the most variability is not only the least productive, but he or she also would have a negative impact on the entire plant because each time he or she stops, the other operators are eventually impacted."

Thomas, who was surprised by the last assertion, needed clarification.

"Daniel, are you telling us that if operator A has an operation time that is more variable than operator B, we can infer that we have a better chance that operator B is the best performer of both?"

"This is exactly what I am saying," replied Daniel.

It was time for a coffee break. "Please take a five-minute break; we will continue to discuss the impact of variability when you come back."

> *When it comes to operator performance, the best performer tends to be the one with the least variability and vice versa.*

After the break, Thomas was very eager to ask other questions. "How do I measure operators' performance? Are the data collected so far enough?"

"This is exactly the next thing I have planned to demonstrate with you," said Daniel.

"How many numbers do I need to characterize each of the two teams' operators' performances?" Thomas, who had some knowledge of statistics, replied, "We normally need two numbers: an average and the deviation." Daniel continued, "That is correct, I am glad you say the average and not a single sample. Without getting into details, in statistics, a single occurrence cannot be used to characterize a distribution—not even the minimum repeatable time generally used by Lean practitioners in kaizen workshops. This is why we use average, which is a single number that typifies a set of numbers. Now, how many types of averages do you know?"

The Industrial Engineering manager jumped in and explained: "I think there are three types: the arithmetic mean (generally called simply "mean"), the median, and the mode. The choice to use any of these three averages depends mostly on the purpose, the distribution of the set numbers, and the objective."

"Correct, Steve. Nothing to add to that," commented Daniel. By experience, he knew that most of the people in factories were not so familiar with the notion of average, mean, median, and mode. He, therefore, went to his favorite and powerful "teach-on-a-simple-example" approach. He took a simple sample to illustrate the definition of each of those notions. "Suppose I draw numbered balls randomly from a bag and obtain the following list: 10, 18, 10, 11, 18, 90, 9, 8, and 18. Let's calculate the minimum repeatable value, the mean, the mode, and the median of this set of values step by step.

"Now, which average are we going to take: the median, the mean, or the mode of the times between parts?" Again, Steve, the Engineering manager, explained that the best one was the mode because that would represent what the operator could do frequently, a sort of performance signature.

Sample	10, 18, 10, 11, 18, 90, 9, 8, 18
Minimum repeatable	10
Mean	$\frac{10+18+10+11+18+90+9+8+18}{9} = 21.3$
Mediane	8, 9, 10, 10, 11 18, 18, 18, 90 Same number of sample left and right
Mode (most frequent value)	18 (it appears 3 times in the sample)

FIGURE 6.1
Computing the minimum repeatable value, the mean, the median and the mode of a set of numbers.

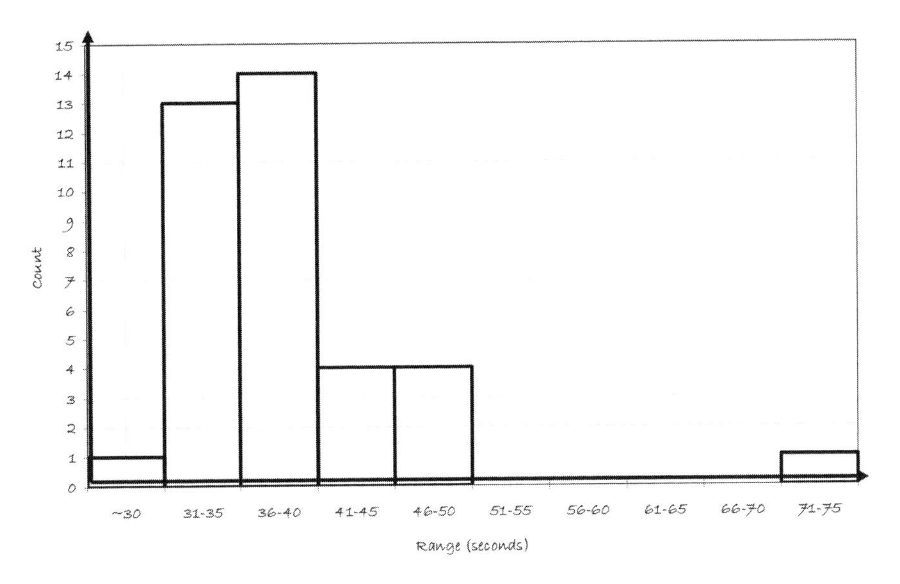

FIGURE 6.2
Histogram of top part times for Group 1.

He also explained that the mean was a very bad indicator because it would include all the breakdowns, which are not due to the operator's work method. This explanation was good enough for Daniel, and he was pleased that it came from a trainee. Therefore, he did not insist a lot on going into more details. He went on and asked both groups to draw a histogram of each t-shirt operator's "top part." As before, he requested they do it manually on another squared poster paper stuck on the wall. This time, each team on a different poster. He also asked the participants to use ranges of five seconds for their histogram bars.

After the drawing (Figures 6.2 and 6.3), Daniel pulled the two teams together and asked what their respective modes were. Group 1's mode was 36 seconds and the mode for Group 2 was 62. He also asked for the mean, the median, and the minimum repeatable. He then commented on the gaps between the numbers, especially between the mean and the mode in the case of Group 2. He explained that this gap meant that Group 2 experienced longer special cause variation than Group 1. "Special cause variations tend to increase the mean, but have no impact on the mode," Daniel insisted. He also explained the bigger gap between mode and minimum repeatable for Group 1 showed even bigger potential for improvement for Group 1 versus Group 2. This meant that not only Group 1 was better, but

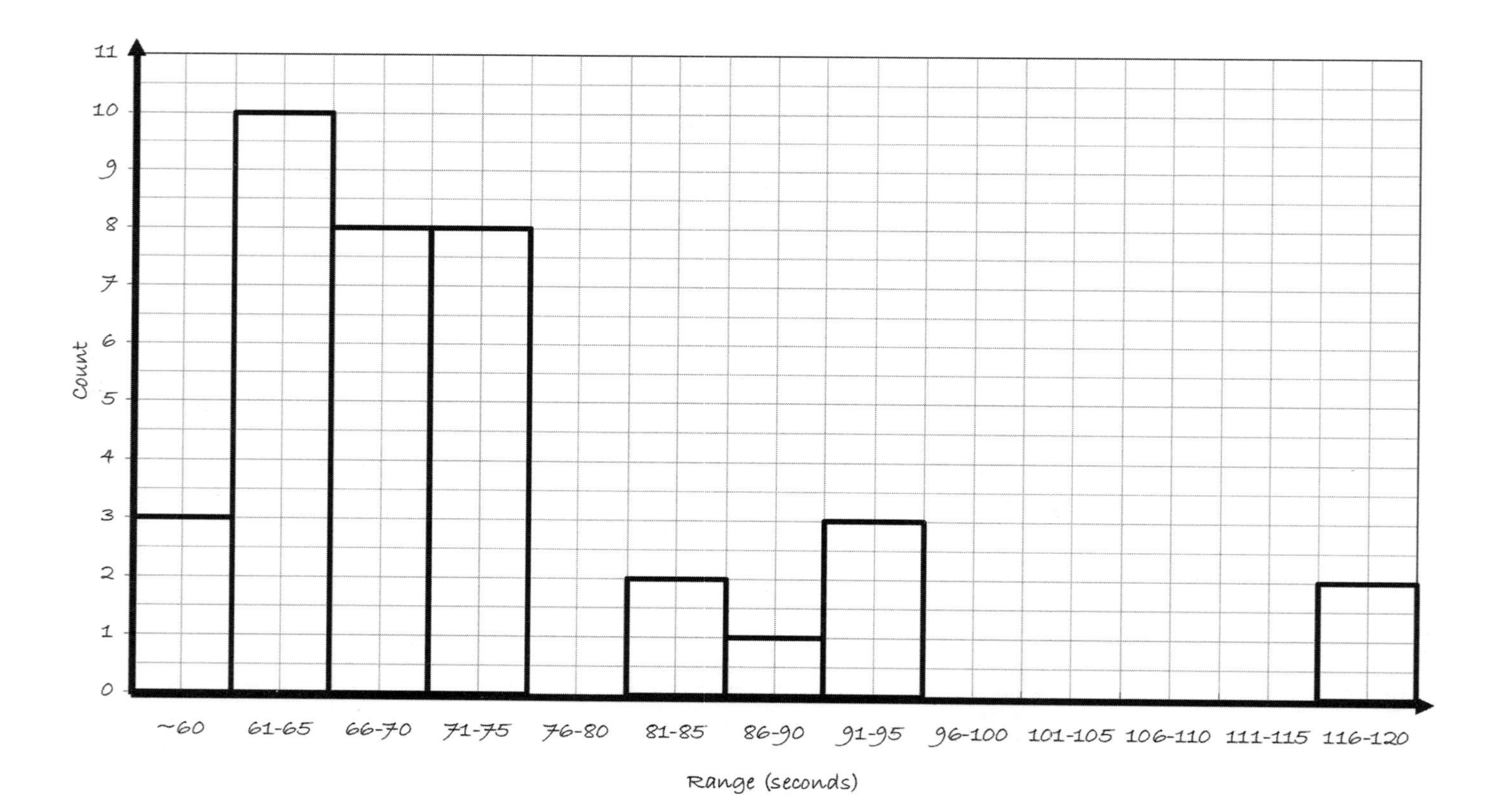

FIGURE 6.3
Histogram of top part times for Group 2.

	Group 1 (seconds)	Group 2 (seconds)
Mode	36	62
Mean	38	73
Median	37	69
Mini repeatable	32	62

FIGURE 6.4
Mode, mean, median and minimum repeatable value for Group 1 and Group 2.

also had more room for improvement (Figure 6.4). Then he concluded, “Before we talk about improvement, let’s focus on variability.”

The mode is the most frequently repeated value. In case of work analysis, it is the most repeated cycle time, a sort of operator’s performance signature.

Special cause variations increase the mean, but have no impact on the mode.

The gap between mode and minimum repeatable shows the potential for improvement regarding operator performance.

The gap between average and mode shows the potential for improvement regarding equipment and availability (breakdowns, changes, etc.).

7

The Variability Index

After the calculation of averages and, especially, the mode, Daniel asked, "Now what's missing to completely characterize each operator performance. Remember, Thomas said that we needed two numbers. What should be the other number?" Thomas explained that this second number should normally measure the variance of the distribution of all the numbers collected. Daniel nodded and explained that although the standard deviation would be the correct indicator, it might be complex to compute, use, and explain on the shop floor. He then proposed a simple method that consisted of counting the number of occurrences in the mode bar and the two adjacent bars, the left and the right to the mode bar. This number of occurrences is then divided by the *total* number of occurrences. He called the result the *Variability Index*.* He did the calculation together for the two groups and found for Group 1: 31/37 = 84%, and 21/37 = 56% for Group 2 (Figures 7.1 and 7.2). He explained that this was completely consistent with the assertion that the operator of Group 1 was the best performer. As predicted, the best performer was the one with the least variation. One of the participants noted that these numbers depended highly on the width of the range, and underscored that for a range of three seconds,† for instance, the variability indices would not have been the same. Daniel offered the following explanation. "You are right. This is a question people have asked me very often. Please let us keep things simple.

* The author used the term *Efficiency* for a variant of this indicator partly published in Patchong, A., T. Lemoine, and G. Kern. 2003. Improving car body production at PSA Peugeot Citroën. Interfaces 33 (1): 36–49. Goodyear and some Japanese companies (including Sumitomo Rubber Industries, Ltd) use the terms Mode Ratio or *Mode rate*. We opted for the term *Variability Index* to underscore that the indicator is variability-related. Also we adopted Japanese companies' computation way, which appeared to be less accurate but simpler and therefore easier to implement on the shop floor than the one originally developed by the author.

† Reminder: The current charts were done with a five-second range.

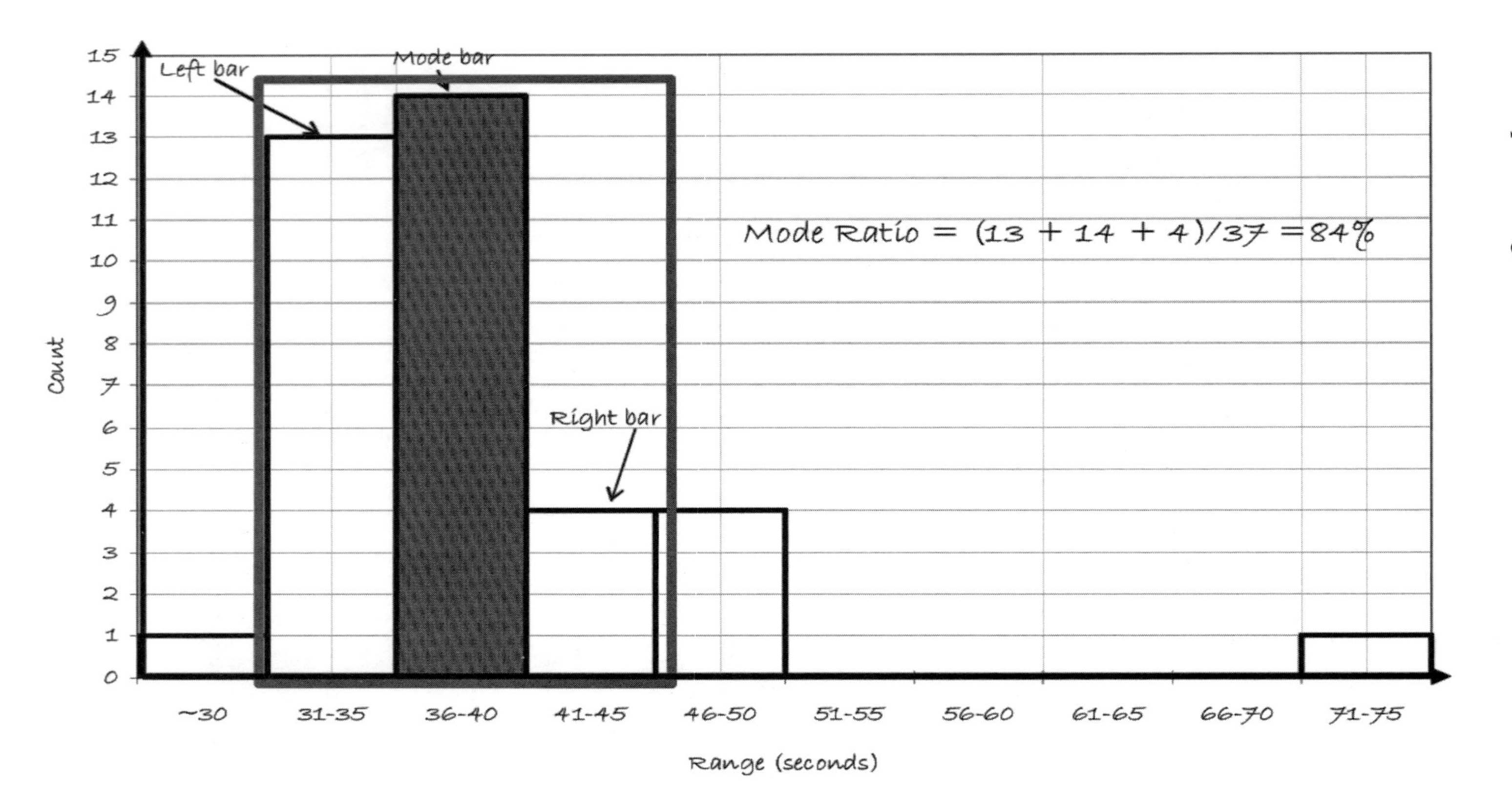

FIGURE 7.1
Computing the Variability Index for Group 1.

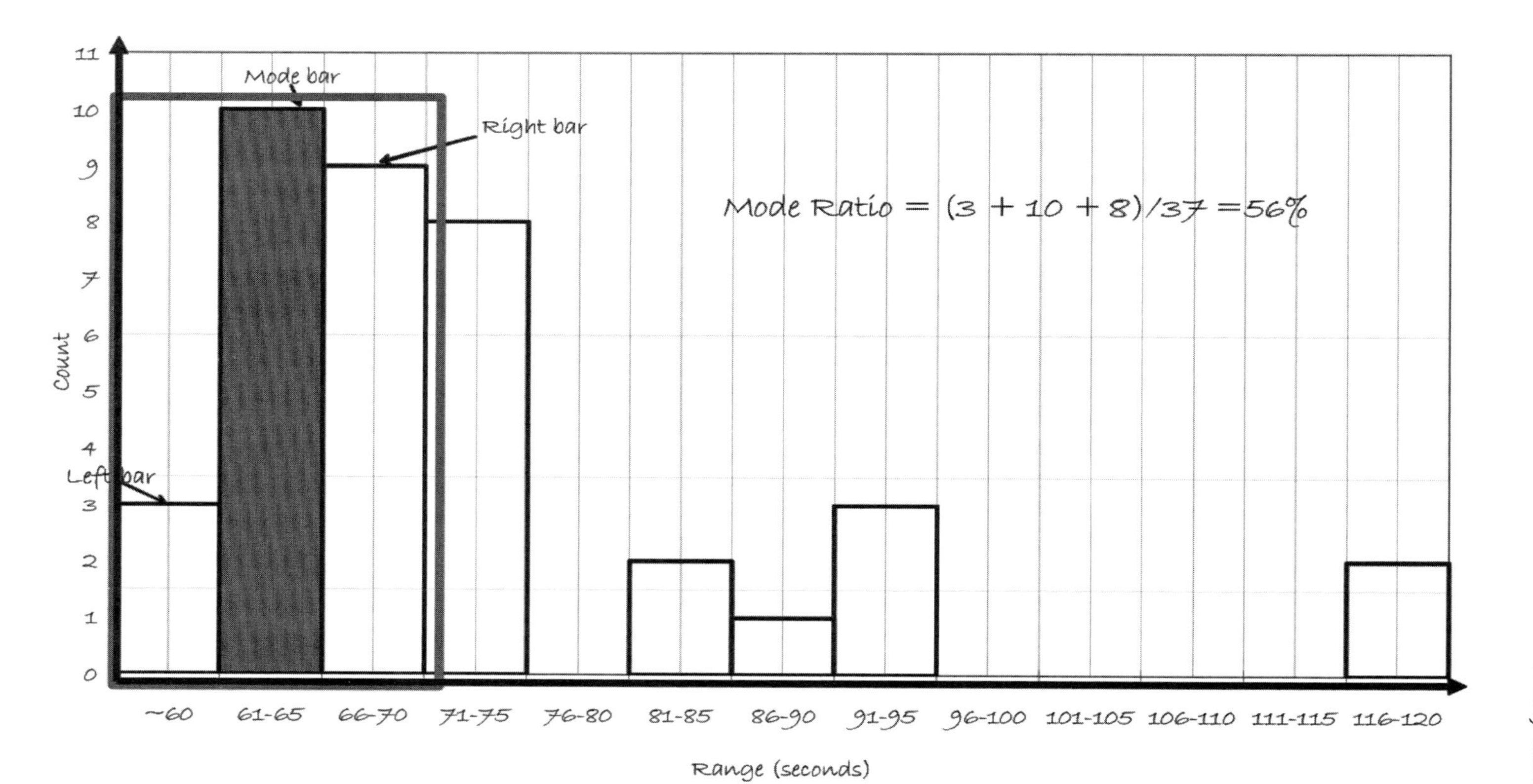

FIGURE 7.2
Computing the Variability Index for Group 2.

The range width may change the Variability Index, but it seldom makes a poor performing operator better than a good one. For the most common cycle time in industry, I mean between 30 seconds and 2 minutes, 5 seconds is a reasonable width. That said, the most important point is that as long as you use the same width for all operators you are comparing, comparisons are valid. An important point in operator performance measurement is consistency of methodology. This is especially true when there is a need to compare or track progress over time. Make sure that you compare apples with apples and oranges with oranges."

Daniel then introduced another indicator derived from the Variability Index that he called the *Adjusted Variability Index*. He explained that the logic of the definition and the calculation was the same as the one of the Variability Index except that the variations due to other causes, not related to the operators (breakdown, changes,...), were removed from the calculation. Daniel then added, "This makes sense especially when we are working on a process that has lot of stability problems; something that will rarely happen if you follow my advice to first achieve minimal machine stability, then address human variability. As we saw in the case of t-shirt folding, those variations, which are special cause variations, are mostly the longest ones. You, therefore, will find the biggest portion of them in the latest bar of your histogram on the right. A good approximate is to subtract from the total occurrences the number of samples in the latest bar of the histogram." After this explanation, Daniel moved toward the flipchart and drew the following table to summarize (Figure 7.3). He then added, "For the sake of simplicity, we will keep the Variability Index throughout this training, but keep in mind that the Adjusted Variability Index is the fairest measurement to assess operators' performance."

> *An important point in operator performance measurement is consistency of methodology, especially when there is a need to compare or to track progress over time. Make sure that you compare apples with apples and oranges with oranges.*

CALCULATING THE RANGE

Before moving to the next part of the training, Daniel thought he needed to come back a little bit to the question he had regarding the range used to compute the Variability Index. He wanted to make sure he had clarified

Variability Index and Adjusted Variability Index

		Denominator: All samples	Numerator: Mode + Left bar + Right bar	Result
Variability Index	Operator 1	37	13 +14 + 4 = 31	84%
	Operator 2	37	3+10 + 8 = 21	56%

		Denominator: All samples – special cause variations (last bar)	Numerator: Mode + Left bar + Right bar	Result
Adjusted Variability Index	Operator 1	37 – 1 = 36	13 +14 + 4 = 31	86%
	Operator 2	37 – 2 = 35	3+10 + 8 = 21	60%

FIGURE 7.3
Computation of the Variability Index and the Adjusted Variability Index for Group 1 and Group 2.

all points and gave the needed tools to everyone to go beyond the training they were receiving today. He, therefore, took a marker and asked the group to join him around the paperboard. Then he started some sketches while talking. "I would like to take the opportunity of the question I had a few minutes ago on the range to explain the precise computation method of the range used to establish histograms as the basis for calculations of the Variability Index. First of all, the rationale supporting the methodology is the 80–20 rule. We want to ensure that 80% of the cycles or the recorded 'top parts' are falling in an interval that represents 20% of the mode value. We also would like the center of this interval to be the mode. As shown on this drawing on the paperboard (Figure 7.4), this means that three ranges (mode bar and the two adjacent bars) should represent 20% of the mode value. From this relationship, one can easily infer what the precise length of the range should be."

Thomas, guided by his practical mindset as always, interrupted Daniel and asked, "Please, Daniel, to make sure that these trainees understand clearly, could we take an example? Say, do it on the results of two groups?" "Perfect! We can do that," Daniel responded. "We will take the example of your two groups." He then flipped the current paper and drew a two-column table, and started a step-by-step running (Figure 7.5). "On the left

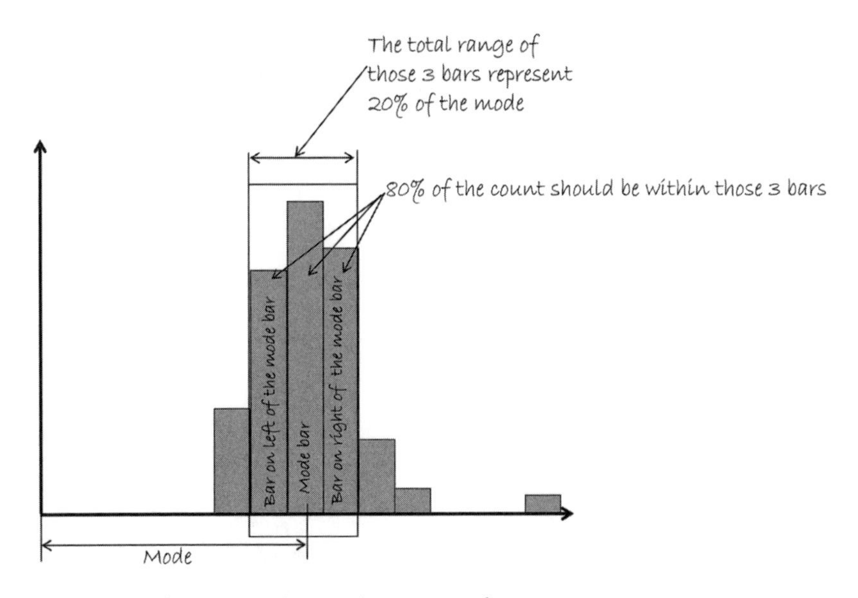

FIGURE 7.4
The methodology is based on two key items: the mode and the rule of 80–20.

column, I will explain the steps. On the right column, I will illustrate with the examples of Group 1 and Group 2. Is that okay?" "Sure," Thomas said.

After drawing the table, Daniel remarked, "Our experience has taught us that it is always suitable to have ranges that are multiples of five." This time without waiting for Thomas's suggestion to use some examples, he proposed: "Let us take the example of the most frequent cycle times you will probably see in the industry." He then started another table while explaining (Figure 7.5). On the first row, he wrote mode values from 0 to 250 seconds, progressing 10 by 10. The second row gives the range as computed with the formula that he had explained previously (Figure 7.6). In the last row, he added the recommended range values, which, except for 1, were all multiples of 5. Then he concluded, "The general rule is that you should choose the closest multiple of 5. There are a few exceptions to

Recommended ranges for common modes

Mode (seconds)	0 - 10	20	30	40	50	60	70	80	90	100	110	120
Range (seconds)	1	2	2	3	4	4	5	6	6	7	8	8
Recommended Range Value (seconds)	1		5									

Mode (seconds)	130	140	150	160	170	180	190	200	210	220	230	240	250
Range (seconds)	9	10	10	11	12	12	13	14	14	15	16	16	17
Recommended Range Value (seconds)	10						15						

FIGURE 7.5
Computation of the range for Group 1 and Group 2.

Description	Example
1/ Compute the mode of the «top parts »	•Mode Group 1 = 36 sec •Mode Group 2 = 62 sec
2/ Knowing that 3 ranges should represent 20% of the mode, compute the precise range	•Range for Group 1 = 36x20%/3 = 2.4 sec •Range for Group 2 = 62x20%/3 = 4.1 sec
3/ Round-up	•Round-up range Group 1 = 3 sec •Round-up range Group 2 = 5 sec
4/ If ranges of the groups are different, choose the biggest number	Max (3 sec, 5 sec) = 5 sec
5/ If range not multiple of 5 chose the Closest multiple of 5 (exception for some cases to be specified)	Closest multiple of 5 sec = 5sec

FIGURE 7.6
Ranges to be used for the most frequent cycle times in the industry.

the rule. Those are ranges that are in double line squares in the table. Now, as I said previously, do not forget that this is not the most important part of the tool. The most important point, if you are to compare, is to keep the same range throughout the comparison." Daniel felt like he had said enough regarding the computation of range. He suggested to the group that they move to the next subject.

8

A-Rank, B-Rank, C-Rank, and D-Rank Operators

Thomas, who was a very practical man, liked the simplicity of the approach used by Daniel to measure operators' variability. He found it very visual and simple to explain and apply. He requested a few minutes to summarize what he had learned. "Daniel, my understanding is that the Group 1 operator can be characterized by his mode and his Variability Index, which are respectively 36 seconds and 84%. The Group 2 operator will be characterized by a mode of 62 seconds and a Variability Index of 59%. Correct?"

Daniel responded, "This is correct and leads to the next point on operator ranking." He then explained to the participants that just like in some education systems, to make things even simpler, operators can be classified into four categories: A, B, C, and D.* He moved to the flipchart to draw Figure 5.1.

As a conclusion of this ranking system, the trainees understood that the Group 1 operator was ranked A, while the Group 2 operator was ranked B.

Thomas asked, "What are we expected to do after the ranking of operator?"

Daniel replied, "Normally your next objective would be to identify the best practices of Operator 1 and train Operator 2 to bring him/her to level A. However, it is not as simple as it sounds. This will be addressed in the coming days.† Now, it is almost 12:30 and, if you do not mind, I would suggest we break for lunch and, when we return, we will go on the shop floor and start working on your most manual area, stage 1 of the Assembly department. My understanding is that this is also your bottleneck area. Correct?" Thomas nodded.

* We adopted the ranking use by SRI (Sumitomo Rubber Industries).

† These points are addressed in the other books of the series dedicated to Standardized Work.

Rank	Variability Index
A	100% - 70%
B	70% - 50%
C	50% - 30%
D	30% - 0%

FIGURE 8.1
Ranks and corresponding Variability Index intervals.

The lunch was quick and the entire group was already back in the room by 1:15 p.m. Before they went on the shop floor, Daniel wanted to move beyond the t-shirt folding simulation and share with them a real example to give them a glimpse of the potential benefits. He always kept in his files some of the successful implementations of the method he had just presented. This was what he called "success stories" and the primary goal of showing this was to create emulation between plants. As he expressed it, "I want people to think 'if they have done it, we should be able to do it as well.'" Daniel displayed the charts and said, "The two charts in front of you are real results from one of our German plants (Figure 8.2). They represent the distribution of time between finished goods as produced by two operators working on the same type of product and the same day. Operator 1 works in shift 1 and Operator 2 works in shift 2. They work on the same machine called MC75 16. As you can see on the charts, Operator 1 delivers 21% more products and his cycle is way more consistent than that of Operator 2. In this example, there is potentially 21% to be gained simply by bringing Operator 2 to the level of Operator 1. This is what I call the hidden treasure. There is virtually no CAPEX (capital expenditures) needed here; you simply need to "copy" one operator's working practices to the other one. This is not expensive, but very rewarding." At this point, a participant interrupted Daniel to ask: "What was the average improvement he had observed when deploying this method in the past?" Daniel explained

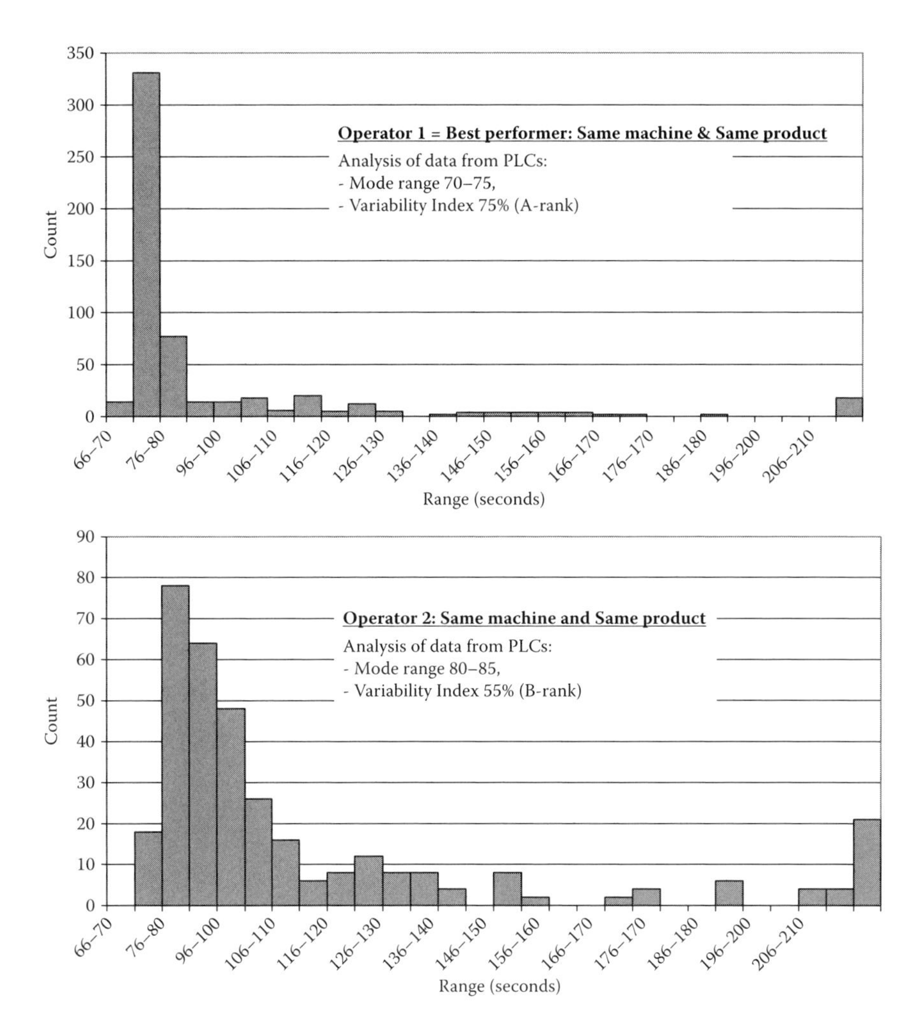

FIGURE 8.2
Real example of two operators making the same product on the same machine.

that the potential gain obviously depended on the initial situation and the effort invested in the deployment. He said, "I have seen gains ranging from 5 to 100%. Any improvement is valuable, especially for an initiative that simply needs people's dedication and energy. Remember: 'Everyone, Everyday, Everywhere.'"

> *Operator Performance Mapping helps identify the best operators very quickly. Then, the next step consists in finding why they are producing more, describing it in a standard, and training all of the others.*

9

Initial Operator Performance Mapping (OPM)

The last charts served as transition for Daniel to explain that the classroom and simulation part of the training to OPM was completed. It was the time for the group to move to the shop floor to perform on-the-floor practice. Next, they took the time to put on their personal protection equipment and headed to the shop floor, directly to the assembly area, which was the bottleneck of the plant. The area had three big workstations around three machines. Daniel had some exchanges with Thomas to better understand the process:

Daniel: I can see that you have three stations here; how many people run those machines, six people on two shifts?

Thomas: No, we run our plant in three shifts to maximize the utilization of our machines, which are very expensive.

Daniel: Does that mean that you have nine people working here?

Thomas: No, we run our machines seven days per week for the same reason. The demand is currently very high.

Daniel: So how many people work here?

Thomas: We have 14 people. We would actually need 15 people, but to give ourselves some flexibility, we have decided not to hire the 15th person. We occasionally use students or temporaries. Let me explain very quickly.

Thomas pulled a piece of paper and a pen from his pocket and explained his calculations (Figure 9.1), then concluded:

Thomas: As you can see, for each job position here, we need to hire five people. Remember, the working time here in France is 35 hours per week; we have holidays, vacation, and some level of absenteeism due to sickness.

Job position staffing

- Plant Working Days: 300 Days/ year
- Plant Working Time: 7200 Hours/ year

- Average working time of 1 operator
  - Weekly working time per Contract: 35 hrs/ Week
  - Yearly working time: 1820 hrs/ year
    - Vacation: 240 hrs/ year
    - National Holidays 60 hrs/ year
    - Sickness: 80 hrs/ year

- Attendance hours of one operator: 1440 hrs/ year

- Needed Operators for 1 Job Fulltime on 300 Days:

$$\frac{7200}{1440} = 5$$

FIGURE 9.1
Computation of the number of workers needed to fill one job position.

Daniel: This means that every time you save a job position, it results in reduction equivalent to five times one worker's compensation. This is really why improving operator performance is key to keeping your plants in high-cost countries.

Thomas: Correct. This will help eliminate bottlenecks in the plant and deliver the requested demand with the same number of people. Remember, I told you we were not delivering the demand today. We have no intention to fire anybody.

Daniel: Well, I would like to underscore that we will do this with all the operators. They need to buy in; if not, we will fail. You also need to keep in mind that for them to buy in, they should see it as a win for them. This will be the situation if the performance improvement actions ultimately lead to better work conditions, and do not require them to work harder but smarter. I will come back to this point when we get into the improvement phase.*

* Workstation improvement is addressed in one of the books of the series dedicated to Standardized Work.

Now, let us go back to performance measurement. You just explained to me that you have 14 permanent operators for three machines. Now, how many types of products do you make on these machines?

Thomas: We produce 23 references on these machines and we do not produce all references on each machine.

Daniel: Do you have a qualification matrix?

Thomas: I am not sure I understand what a qualification matrix is.

Daniel: Well, do you have a document where the level of qualification of each of your workers is recorded on all existing machine-product combinations?

Thomas: I am not sure.

Daniel: Okay, let me draw it somewhere.

Daniel pulled the group into a nearby office where he could find a flipchart and drew the matrix shown below based on all information he received from the participants (Figure 9.2). Every time an operator could run one of the three machines to make one of the 23 references, he drew an "X" in the cell. He then ended up with 47 cells with an "X". That is what he called the first step of the mapping process: knowing exactly who can produce what, with which machine. The next step he specified was performance measurement of those 47 configurations of the triplet: operator, product, and machine. That meant, just as they did in the classroom, measuring the mode range and the Variability Index of each configuration. This initial mapping, insisted Daniel, would give a good overview of the current situation. He also explained that this was actually the easiest way to do an initial Operator Performance Mapping (OPM).

Thomas took a marker, went to a flipchart, and drew a table while explaining (Figure 9.3), "Daniel, let me summarize this to make sure that I have understood this. We are trying to know the performance of all 14 operators in all possible situations. The operator's machine is the primary determinate of the situation. Because the three machines are different, an operator's performance may be different from one machine to another. Correct?" Daniel nodded. Then Thomas continued, "The performance of each operator also will depend on the reference of the product he is working on. This is because the work operation may be different from one product to another. So, at the

Product→	Machine 1									Machine 2							Machine 3						
	1	2	3	4	5	6	7	8	9	10	11	12	13	14	15	16	17	18	19	20	21	22	23
Ed	X		X				X	X	X									X	X				
Rob		X		X			X		X												X		
Tim					X		X			X			X										
Elen			X				X																
Ali		X				X																	
Jodi									X	X							X	X	X				
Stan							X				X							X					
Peter									X														
Alex												X		X		X							
Joe										X				X									
Al									X		X				X								
Dan								X										X	X	X	X	X	
Bill																		X				X	X
John																						X	

FIGURE 9.2
Basic qualification matrix that shows who can make which product on which machine.

Number of operators	Number of machines	Number of products	Operator Performance Mapping size
14	3	23	47
14X3X23=966 possibilities			

FIGURE 9.3
Computation of the number of workers needed to fill one job position.

end of the day, the theoretical number of possibilities* is: 14 × 3 × 23 = 966, but only a few of these possibilities are feasible or operated in practice. These are the configurations we are looking for. These are the 47 cells where you drew an X. Correct?" Daniel replied, "Your summary is absolutely perfect; no need to add anything to that."

At this stage, Thomas, who was already thinking about the next step, asked the following question: "Daniel, how many samples of 'top part'† are we going to need to measure the performance of each configuration?" Daniel explained that a shift-long record would be representative enough of the performance of the 14 operators. "A shift-long. That is a lot to record with stopwatches," said Thomas. "Oh, yes. The best way to collect those numbers is definitely not the stopwatch. The 'top part' is something you easily will find in any PLC.‡ Even when there are no PLCs, which is not your case here, it is very easy to install a detector that will catch a finish, an arrival, or movement, and send the signal somewhere to be recorded," Daniel responded.

He continued: "Going back to your initial question on the number of samples needed, I would say that in statistics, the general rule is that the more samples you have, the better your analysis. Accuracy increases with the number of samples; that is a fact. The mode is sometimes very easy to find. The mode of a manual operation could be identified with only a few samples. Sometimes even 10 cycles are enough to find a mode. The Variability Index is more sensitive as its accuracy will increase with the

* Fourteen is the number of operators, 3 is the number of machines, and 23 is the number of product references.

† Time recorded every time a part is produced.

‡ Programmable Logic Controller.

Shift pattern

	Morning	Afternoon	Night
Monday	Shift B	Shift C	Shift D
Tuesday	Shift A	Shift B	Shift C
Wednesday	Shift A	Shift B	Shift C
Thursday	Shift D	Shift A	Shift B
Friday	Shift D	Shift A	Shift B
Saturday	Shift C	Shift D	Shift A
Sunday	Shift C	Shift D	Shift A

FIGURE 9.4
The plant 4 × 8 shift pattern -4 shifts and one shift every 8 hours.

number of samples. Again, the most important thing here is consistency in the rules. The rules should stay the same for any comparisons to be valid."

Daniel asked John, the engineering person, what it would take to have access to the data. He quickly realized that the engineer was more of a procrastinator than a "do-it-now" person. Therefore, he decided to press him a little bit. He explained that he should try harder given the exceptional circumstances, and that worked. They made the decision to undertake the needed modifications in the afternoon so that the next day all of the data would be available. John promised that tomorrow they would have data for the previous 10 days. Daniel was happy because this would allow the team to develop a fully automated system by the end of the week. But, as always, he wanted the team to do things manually first. As previously, he insisted that this was the best way to understand a new method. He then turned to face the group and said, "As we saw previously, you have 14 operators running the 3 machines. We would need to perform measurement for all of these 14 operators. What is your current shift pattern?"

In response to this question, Thomas grabbed a paperboard and wrote down the shift pattern as shown in Figure 9.4.

After Thomas's explanations, Daniel continued: "Well, we have 12* operators scheduled from now to tomorrow morning. We need to measure their performance. One hundred 'top parts' per operator will be enough

* Three operators per shift times four shifts.

for the purpose of this training. I suggest we organize ourselves into three groups to observe the three operators of the first morning shift (Shift B) from now to the end of the shift at 2:30. Please take 100 consecutive 'top part' times per operator. Then, the same groups will observe the operators of the afternoon shift (Shift C). You should also take 100 consecutive 'top parts.' Tomorrow, we will do the same for Shift A, which works in the morning."

Six of the most enthusiastic participants volunteered to come back to the plant that evening to collect the data of Shift D's operators. That made only 12 of the 14 operators working on the three assembly machines. "Now," Daniel asked, "What about the two remaining operators?" The production manager explained that they were working on the Tuesday morning shift, but in a different process area taking care of other machines. He then offered to have them back on the assembly machines on Tuesday morning to run the machine for just the time needed to collect enough data. That would certainly create some disruptions in production, but he assured it was not a big deal and insisted he would be able to handle the situation.

Now that the data collection was organized, Daniel handed out a few papers to each group on which to collect the data. He underscored that this was an opportunity for the participants to observe intensively the work of operators. He, therefore, encouraged them to take advantage of the situation to identify any simple, quick, and zero-CAPEX improvements.

After the collection of data of Shift B and Shift C, Daniel released the team, as he knew that some of them would come back again in the evening to collect additional data. Daniel said, "That's it for today. I will see you this evening in the assembly area at 10 p.m., or tomorrow at 8 a.m., same place. Have a nice evening."

At 10 p.m., all of the trainees who volunteered to come were in the assembly area. The data collection process took place without any major problem. After 100 "top part" measurements, everyone went home, a little bit exhausted from the very long day. Driving back to his hotel, Daniel thought about an old friend who always told him that people better remember things they learned the hard way. He thought, "May this exhaustion be the cement that will make today's learning stick in their heads." It had been a tough day for Daniel as well and he had no trouble falling into a deep sleep.

Tuesday was as sunny as Monday. The entire group arrived roughly 10 minutes before the beginning of the session. Before starting, Daniel made a quick summary of the previous day's activities. He systematically asked all participants what they had learned on Monday. It appeared that, in general, the trainees were really excited about the previous day. Their comments clearly indicated that all participants understood why variability was their biggest enemy and how they could measure it. They were impatient to get the results of the data collection of the remaining shift in order to make an analysis. After this quick summary, Daniel suggested that the group go directly to the shop floor to collect data from Shift A and the two remaining operators. At 10:30 a.m., the group was back in the room. They now had data from all 14 operators. At this very moment, John, the engineering person, announced that he was able to collect data from the PLCs. All the data from the previous 10 days of production were in a memory stick. The group now had more data than needed and was ready to move to the next stage. Daniel jokingly told the group: "Now you can use your computers." The group was divided into four subgroups of two people each, and every group needed to process roughly 12 sets of data. They knew they had to start with the two following tasks:

Task one: Draw the histograms with a bar range of 5 seconds for each of the 47 configurations.

Task two: Find the mode bar for each of the 47 configurations and write them in the qualification matrix in place of the X.

The group delivered Figure 9.5.[*] All mode bar range results were written in a big matrix displayed on the wall.

Daniel congratulated the group and asked them to proceed to the next step and perform the following two tasks:

Task three: Calculate the Variability Index, just as was done together previously.

Task four: Draw a histogram chart with the Variability Index on the *x*-axis: minimum 0% and maximum 100% and a range of 5%. Superimpose all the configurations with the same Variability Index in the same column.

Daniel explained that the result of the last task would help give a good overall assessment of operators' performance. The group drew Figure 9.6.[†]

---

[*] Such automation.

[†] To the best of our knowledge, some Japanese companies (including SRI) use a similar chart.

	Machine 1 (Mode range - in seconds)									Machine 2 (Mode range - in seconds)							Machine 3 (Mode range - in seconds)						
Product→	1	2	3	4	5	6	7	8	9	10	11	12	13	14	15	16	17	18	19	20	21	22	23
Ed	71-75		66-70				71-75	81-85	71-75									76-80	76-80				
Rob		66-70		61-65			71-75		71-75												81-85		
Tim					66-70		71-75			71-75			71-75										
Elen			66-70				66-70																
Ali						71-75																	
Jodi									71-75	71-75							66-70	76-80	76-80				
Stan							71-75				71-75							76-80					
Peter									71-75														
Alex												61-65		71-75		71-75							
Joe										71-75				71-75									
Al									86-90		81-85				71-75								
Dan								86-90										76-80	71-75	71-75	71-75	61-65	
Bill																		76-80				76-80	76-80
John																						66-70	

FIGURE 9.5
Result of Task two -47 mode bar range values.

Variability index														
					Bill Prod 18 76-80									
					Dan Prod 21 71-75									
					Joe Prod 10 71-75									
					Alex Prod 12 61-65									
					Stan Pro 11 71-75									
					Tim Prod 10 71-75									
				Jodi Prod 10 71-75	Ed Prod 18 76-80	Joe Prod 14 71-75								
				Stan Prod 18 76-80	Jodi Prod 18 76-80	Tim Prod 13 71-75	Al Prod 15 71-75							
		Jodi Prod 17 66-70	Dan Prod 20 71-75	Tim Prod 7 71-75	Stan Prod 7 71-75	John Prod 22 66-70	Al Prod 11 81-85							
	Tim Prod 5 66-70	Elen Prod 3 66-70	Dan Prod 19 71-75	Rob Prod 7 71-75	Jodi Prod 9 71-75	Dan Prod 22 61-65	Alex Prod 16 71-75							
	Rob Prod 2 66-70	Rob Prod 4 61-65	Dan Prod 18 76-80	Ed Prod 7 71-75	Elen Prod 7 66-70	Bill Prod 22 76-80	Alex Prod 14 71-75	Jodi Prod 19 76-80				Ed Prod 14 81-85		
	Ed Prod 1 71-75	Ed Prod 3 66-70	Ali Prod 6 71-75	Ed Prod 8 81-85	Peter Prod 9 71-75	Ed Prod 9 71-75	Rob Prod 9 71-75	Ed Prod 19 76-80		Rob Prod 21 81-85	Dan Prod 8 86-90	Al Prod 9 86-90		
100% ~ 80%	79% - 75%	74% - 70%	69% - 65%	64% - 60%	59% - 55%	54% - 50%	49% - 45%	44% - 40%	39% - 35%	34% - 30%	29% - 25%	24% - 20%	19% - 15%	14% ~ 0%
A-rank			B-rank				C-rank				D-rank			

Legend

Operator → Ed; Product type → Prod 14; 81-85

Mode bar range in seconds
(Ed's mode is between 81 and 85 seconds)

Machine 1

Machine 2

Machine 3

FIGURE 9.6
Initial operators' Variability Index distribution.

Daniel checked the latest chart and commented on the results. "As you may see, this chart gives the overall distribution of the work performance in the assembly area. At a glance, you can quickly get a clear idea of where your operators stand. For instance, on this chart, you will notice that the average Variability Index of your operators is somewhere between 55 and 59%. This is a good indicator of the overall performance of all operators. To finish this OPM, below are the two last tasks you should carry out."

Task five: Calculate the proportion of operators ranked A, B, C, and D.
Task six: Calculate the overall average mode of the all operators.

Using the Variability Index of all configurations, the group was able to calculate the proportion of configurations in each raking: A, B, C, and D. They came up with Figure 9.7.

Using the same Variability Index numbers, they also were able to compute the average Variability Index of all configurations and found it to be 55%. This meant that the current overall work performance level in the assembly area was at a low "B" level. That surprised the participants, as they expected the number to be much higher.

"What level did you anticipate?" Daniel was interested in getting their input. "I hear different answers between 75 and 85%, but what I hear most frequently is 80%. We are not at this level today, but let's make it an objective," he suggested.

Daniel explained that the number was not so bad and encouraged them to move to the next step by challenging themselves to reach an average

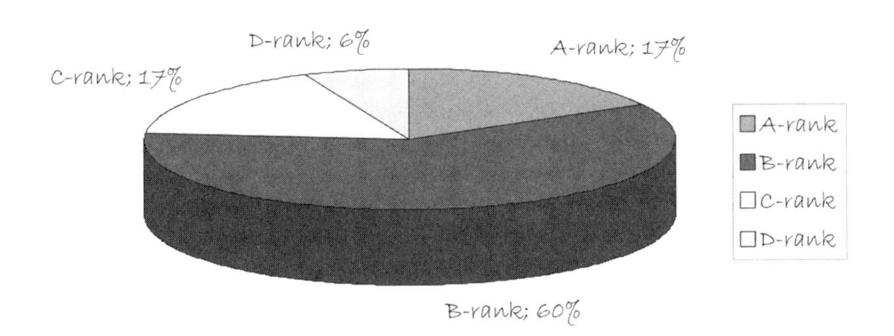

FIGURE 9.7
Initial operators ranking -percentage of As, Bs, Cs, and Ds.

Variability Index level of 80% by the end of the year. He then added, "This is what you expected to get today. A Variability Index of 80% will be your goal. By the way, as I explained earlier, this objective goes very well with the Pareto rule of 80–20, which is the rationale supporting the measurement method. Let us make 80% your variability index objective." Thomas suggested that they call this endeavor "The Quest for Mode 80." All participants agreed.

This is where the group concluded the initial Operator Performance Mapping training.

10

Detailed Standardized Work Deployment Steps[*]

The end of the second day of training was quickly approaching. This was also the end of the training module dedicated to the Operator Performance Mapping (OPM) training.

A small number of participants came to Daniel and questioned him about the next steps. Daniel took the opportunity to pull the group together to summarize and discuss what came next. "This has been a very good day. You have worked very hard and achieved unexpected results. Before I let you go for the evening, I would like to give you some perspective about the next steps."

Daniel advanced to the flipchart and asked the trainees to follow. He grabbed a marker and said, "You now have a clear view of the performance of each operator per machine[†] and per product. You know your best performers and the ranking of your operators from A to D. Again, as you all know, this is what we call Operator Performance Mapping. At this point, it is complete. As I underscored previously, the OPM[‡] is just a tool for Standardized Work deployment. We are only at the beginning of Standardized Work deployment, which should be performed in the following four steps as I told you yesterday while introducing the training. If you remember well, I also told you yesterday that I would give you more details regarding those four steps.[§] And, here you are."

* This session gives an overview of the five steps of Standardized Work deployment.

† *Machine* is used here as a generic term to characterize a workstation.

‡ Operator Performance Mapping.

§ Operator Performance Mapping, which is the subject of this book, is one of the tools of Standardized Work deployment. The other remaining tools of the four steps are addressed in other books of The One Day Expert series dedicated to Standardized Work.

Thomas displayed the chart showing the four steps of the Standardized Work deployment (Figure 10.1).

Step 1: Capturing the current state

In this step, two sets of tools are used. The first set is comprised of the Operator Performance Mapping we just implemented. The second set of tools are the Standardized Work support documents. The first one is Process Analysis. It is used very early in the deployment process to estimate existing capacity and identify the main causes of capacity losses. At this stage, it is possible to get a fair estimate of the extent of special main cause variation and common cause variation. After this stage, if the cause of the problem is special, then you should focus first on fixing the special circumstances before addressing operators' work improvement (common cause variation). However, you do not need to achieve 100% stability. If your process is stable enough, then you will need to write and implement the rest of the forms, which include standardized work chart, standardized work combination table, and operator work instructions.

Step 2: Improving the process

Here you will implement very quick improvements based on shop floor observations. You will look for easy ways and small equipment to improve three things: operator work methods, operator work conditions, and equipment availability. The heart of this improvement step is a series of long and intensive observations called *tachinbo.*[*] This is Japanese for standing still for long hours at the same location. As you will see when we practice, it is all about seeing, and learning to see, simple opportunities for improvement. It is a very powerful development tool for those who are starting to implement Standardized Work, or more generally, the Lean journey. You are probably asking yourselves: What is this 'Sharing the black[†] book stuff?' Along with *tachinbo,* this is the other tool used to collect improvement ideas.

[*] This tool is detailed in proceeding books of The One-Day Expert series dedicated to Standardized Work.

[†] This tool is detailed in proceeding books of The One-Day Expert series dedicated to Standardized Work.

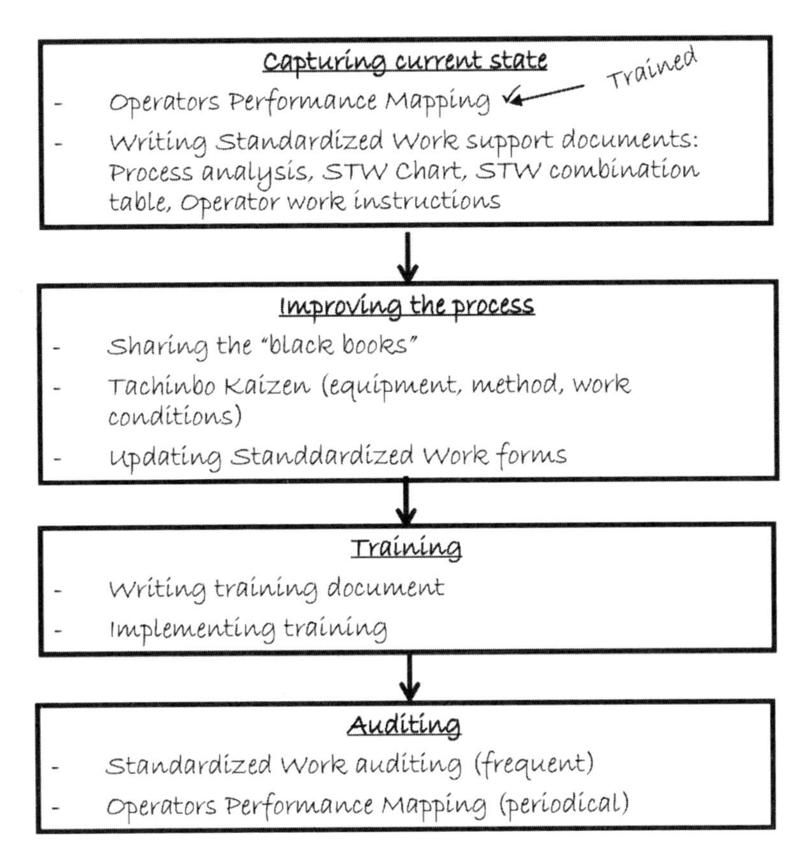

FIGURE 10.1
Details of the Standardized Work deployment.

The approach here is to observe every step of each operator doing the same job, and identify all "best practices" for each of them. Each operator's set of best practices and ideas for improvement is what we call his or her "black book." Therefore, "Sharing the black books" means that all those best practices or "secrets" are shared and included in the new final Standardized Work method to be learned and applied by all operators. Please note the huge role played by operators in this step. This step ends with updating all standardized documents developed in the first step to include all the changes made during the improvement step.

Step 3: Training

It is not enough to improve and write standardized work documents. It is also very important to train operators to use these new methods. You should not expect people to apply new work methods if they are not trained. As you will see by the end of this week, writing job instruction documents is as important as using the right method to train. The right method is the one that allows operators to get to the highest level of performance the most quickly. I will give you more details on Friday.*

Step 4: Application and auditing

This is the last and the most important point. After having carried out all the four previous steps, Standardized Work deployment is only worth starting if there is a real application. In order to make sure that Standardized Work is applied at every station, frequent auditing is needed. For the auditing to be frequent, the auditing process needs to be simple. The key words here are "simple" and "frequent." What does "frequently" mean? Well, if your system is manual, the whole process is time consuming. Therefore I would suggest you check every month to every quarter. Now if the whole thing is automated†, you can monitor on a weekly or even a daily basis. A daily check could be time consuming and would provide very little information. I therefore advise you stick to a weekly monitoring. Besides standardized work auditing at workstations, you will have to use the OPM periodically (every week if it is automated, or every quarter/semester if it's manual) to monitor the progress of operators and set your training goals. In order to perpetuate continuous improvement of existing work methods, please remember that auditing is the only way to see problems and solve them.

"As you can see on the chart I just drew (Figure 10.1), we are at the very beginning of the process," Daniel remarked. "We have only done the initial OPM." While drawing another table (Figure 10.2), Daniel said, "We will continue tomorrow with Process Analysis and the

* Training is addressed in one of the books of the One-Day Expert series dedicated to Standardized Work.

† Such automation were implemented at Goodyear plants.

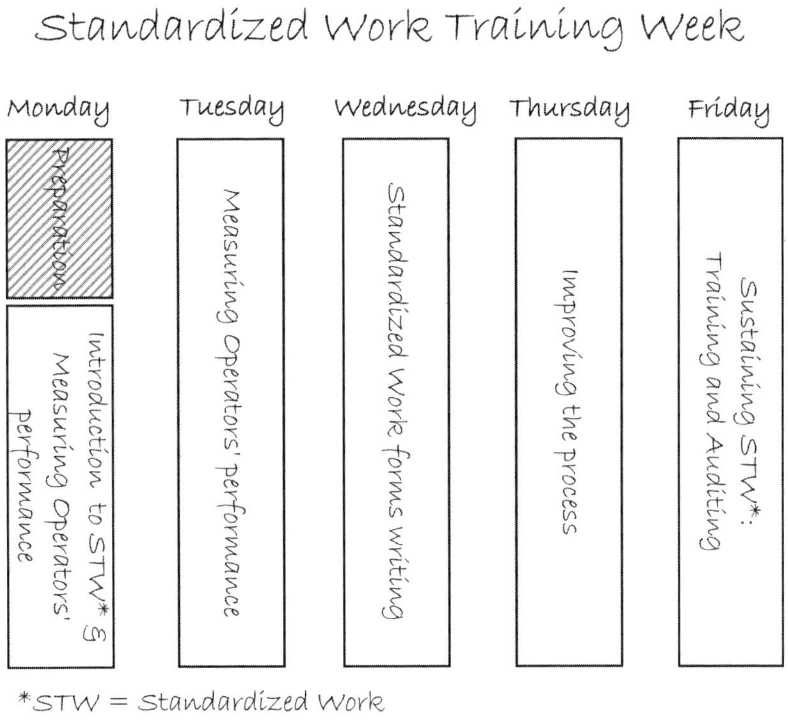

Standardized Work Training Week

Monday
Preparation
Introduction to STW* & Measuring Operators' performance

Tuesday
Measuring Operators' performance

Wednesday
Standardized Work forms writing

Thursday
Improving the Process

Friday
Sustaining STW*: Training and Auditing

*STW = Standardized Work

FIGURE 10.2
Standardized Work training week schedule.

writing of the three other Standardized Work documents: standardized work chart, standardized work combination table, and operator work instructions. Then, Thursday will be dedicated to Process Improvement. We will conclude the training on Friday afternoon with two key activities needed to sustain Standardized Work: training and auditing.* You will see that these are very easy tools that you will learn to understand and practice in only one day. These tools will help you improve the performance of your operators and, when I return here next time, I will see how far you are from your objective of bringing your operators to an average Variability Index of 80%. I can tell you that this is a big challenge and I completely trust that you ultimately will be able to reach it. I do not know if it will happen the next time I visit, but I am sure that you will make it happen."

* Each part or the training mentioned in the text and the table corresponds to a book of The One-Day Expert series dedicated to Standardized Work.

The training regarding operators' performance measurement ends here. Thomas thanked Daniel for the training module on OPM and promised that they will achieve great progress by the next time he visits. Daniel said, "I will let you go home now. See you tomorrow. We will be talking about Standardized Work document writing, which includes Process Analysis, Standardized Work Chart, Standardized Work Combination Table, and Operator Work Instructions."

The heart of the Standardized Work improvement process step is a series of long and intensive observations called tachinbo.

No one should be expected to know the job if he or she has not been trained properly.

Any Standardized Work deployment without auditing is a huge waste of time and is not worth starting.

11

Epilogue: The Quest for Mode 80

Three months later, Thomas invited Daniel to visit his plant. It was the beginning of autumn, but the weather was still quite warm and sunny. It reminded Daniel of the same beautiful day when he conducted his training.

The plant director left his office to meet Daniel downstairs in the lobby of the main building.

Good morning, Daniel. I hope you had a good trip. If you are not tired, I would like to take you directly to the shop floor, but today I will be asking you questions, explained Thomas with a smile.

I am not sure I will know all the answers, replied Daniel, smiling, too.

When they walked along the machines, it was obvious that the plant had undergone a big change. The takeaway conveyor was filled with products and, surprisingly, all of them were equally distributed. This was a visible sign that the variability of the process had gone down.

Thomas: Daniel, tell me what you think of what you see.

Daniel: Oh, my friend, I see you have worked a lot and made great progress with your people. Here, at the end of the production line, I see results and I am sure you have achieved a lot of the things we talked about during our one-week training.

Thomas: Operators Performance Mapping (OPM) was a great thing to start with, then we conducted capacity analysis of all our lines. This analysis opened our eyes and we saw how much variability we had in our process. I have to admit, at the beginning we struggled a bit.

Daniel: Why? Didn't you conduct any tachinbo? By the way, do you still remember what it is?

Thomas: Sure, Daniel, it became a regular part of our improvement activities, but, to be honest, we had a problem getting employees involved in making process observations. It was something completely new for our culture, but, at the same time, it helped to change it. And, you know how we selected people?

Daniel: You asked for volunteers? (Daniel was interested in Thomas' approach.)

Thomas: Not really, it was the first time in my career that I took a very traditional approach. I told my people that everyone would have to make one tachinbo, no excuses. They were resistant at the beginning, but the first results exceeded both our expectations.

Daniel: And, it was what helped you to change this plant.

Thomas: That's right, but it was not the end of the story.

Daniel: It seems you haven't forgotten our workshop and its key elements.

Thomas: I am a very good student. It was important to standardize new things, first through solid training. People were not accustomed to this approach. Many of them thought that training was just to tell them what to do, while I wanted them to be more practical. Moreover, they were very resistant to each kind of follow-up or audit. Nevertheless, all the pieces of the Standardized Work concept worked out very well for us, and our employees have been very committed to it. I think we may need more time to talk about it. Let's make it during dinner tonight. For the time being, I would like to talk about our achievement concerning operators' variability. After taking care of equipment variability, by reducing the level of breakdowns, we decided to take care of our operators by training them in the best methods we describe in our Standardized Work forms. Thereafter, we did another OPM and measured the Variability Index. As you could see on the chart, we have not reached the 80%, which was our target, but we improved a lot (Figures 11.1 and 11.2). Operators' overall Variability Index increased by 10 points and the production followed as well. You may remember that when you visited us last time we were producing 3015 parts per day. Now our total production has increased by 13% to reach 3405 parts per day (Figure 11.3). This is really amazing. Especially so when you consider we spent virtually nothing to make this progress. I could not have imagined that it was possible to achieve that much without any investment. This impressive progress has been a key point in the buy-in of my people here. They really

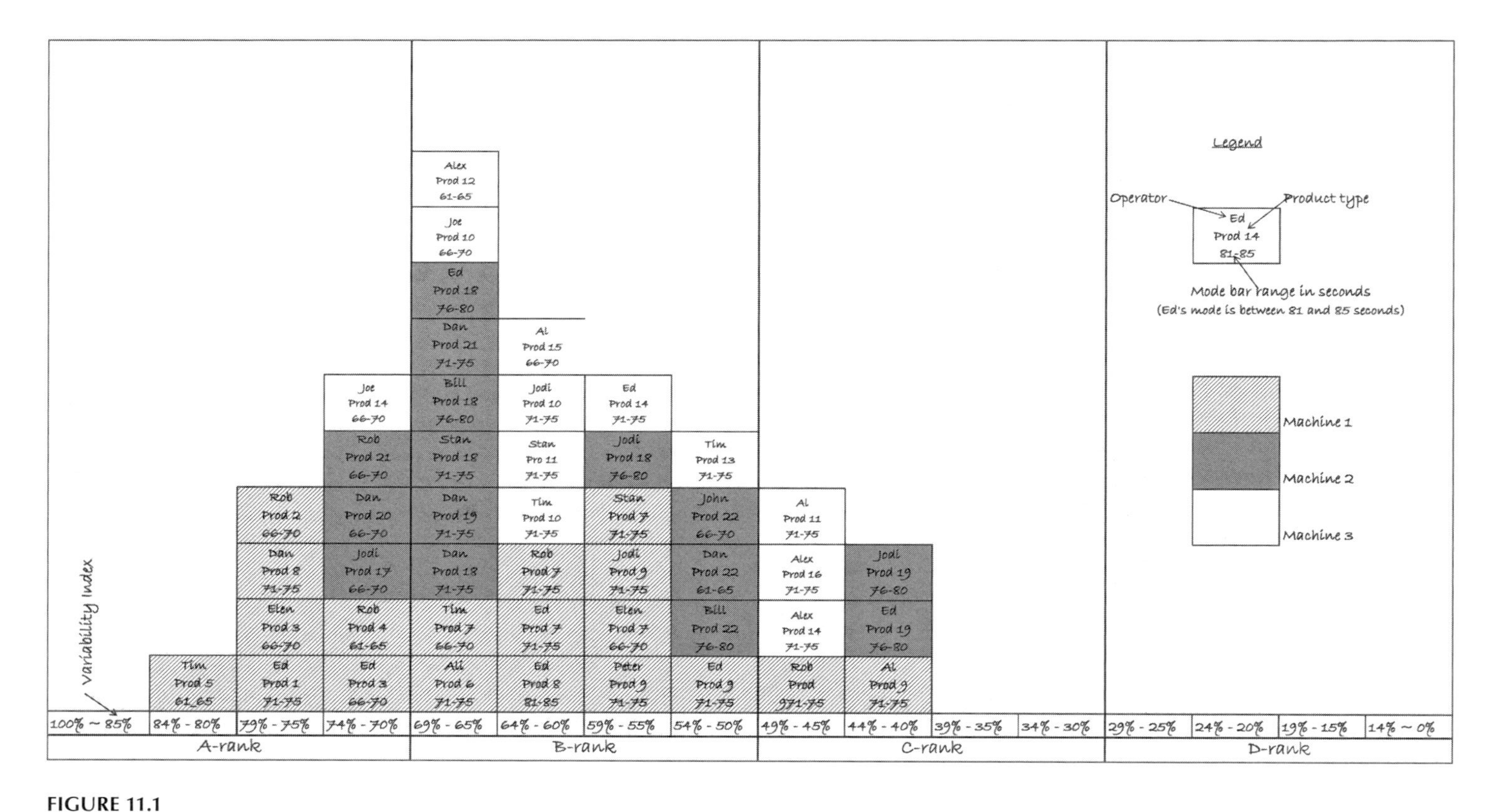

FIGURE 11.1
Final operators' Variability Index distribution.

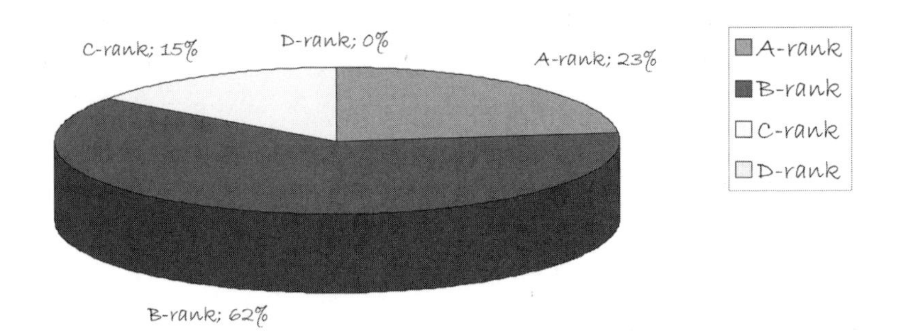

FIGURE 11.2
Final operators ranking -percentage of As, Bs, Cs, and Ds.

> saw the reward of their efforts. They now see that by working smarter, not harder, they are able to produce more. Ergonomics has improved and nobody has been fired. Our morale is definitely up.

Although they did not have the same background, Thomas and Daniel shared the same passion for changes made on the shop floor. On one side, Thomas was visibly excited to tell his story and show the results on the shop floor. On the other side, Daniel was equally thrilled to listen and ask questions. Thomas suddenly noticed it was getting late, so he suggested they continue their discussion over dinner. Daniel was hungry and readily agreed.

Thomas and Daniel drove to a Mexican restaurant downtown. They had a very long dinner. The food was good and spicy, as in most Mexican restaurants. Thomas shared the entire story of what happened in the plant after the Standardized Work workshop. Their achievement went beyond the results of the OPM because they also deployed other tools of the Standardized Work's five steps.* As he stated it: "This has not been a catwalk; we went across some serious roadblocks, but were able to overcome, thanks to the team motivation." He also explained that throughout the process he always tried to adapt his leadership style to each of his team members' development and readiness: directing, supporting, coaching,

* Details of results concerning the implementation of those five steps are recounted in proceeding books of the Standardized Work series.

Scorecard

	July 2012	September 2012	Progress
Average Variability Index	55%	65%	+ 10 points
Production	3015 parts / day	3405 parts / day	+13%

FIGURE 11.3
Improvement scorecard.

and delegating. He was very used to this customized leadership since the training he took a few years ago on Situational Leadership.* He also insisted that the other point of focus for him was to make sure that people would get some "quick wins" along the way to instigate and nurture the motivation flame. The biggest threat, he underscored, was resistance to change.

"Daniel as you well know, change is not easy and people tend to be resistant. I knew that from my previous experience before I came here. The stiffness of resistance was higher than what I anticipated. This was, at least in part, due to the recent history of numerous management changes. People had seen a lot of plant managers coming and going with no constancy in their messages. Therefore, they thought 'here we go again, another new plant manager with new stuff just like his predecessors; let's wait for the next one.' This disillusion was the toughest thing to conquer. The early successes we clinched were instrumental. People started to believe again and the level of cynicism among the team substantially decreased." As you have seen on the shop floor during our visit, we have got a more ambitious target now, which is that a year from now 80% of our workers should be ranked A, 20% of them ranked B. It means that we will have no more Cs and Ds (Figure 11.4). The team is with me and we are confident that we can meet this target.

Thomas paused for a few seconds and added, "Daniel, as you can see, we have done a lot, but I think there is even more difficult work ahead of us—keeping these standards and continuing to improve."

* Situational Leadership is a leadership theory developed by Paul Hersey and Ken Blanchard. More details can be obtained in Hersey, P., K. H. Blanchard, and D. E. Johnson. 2007. *Management of organizational behavior: Leading human resources.* Upper Saddle River, NJ: Prentice Hall.

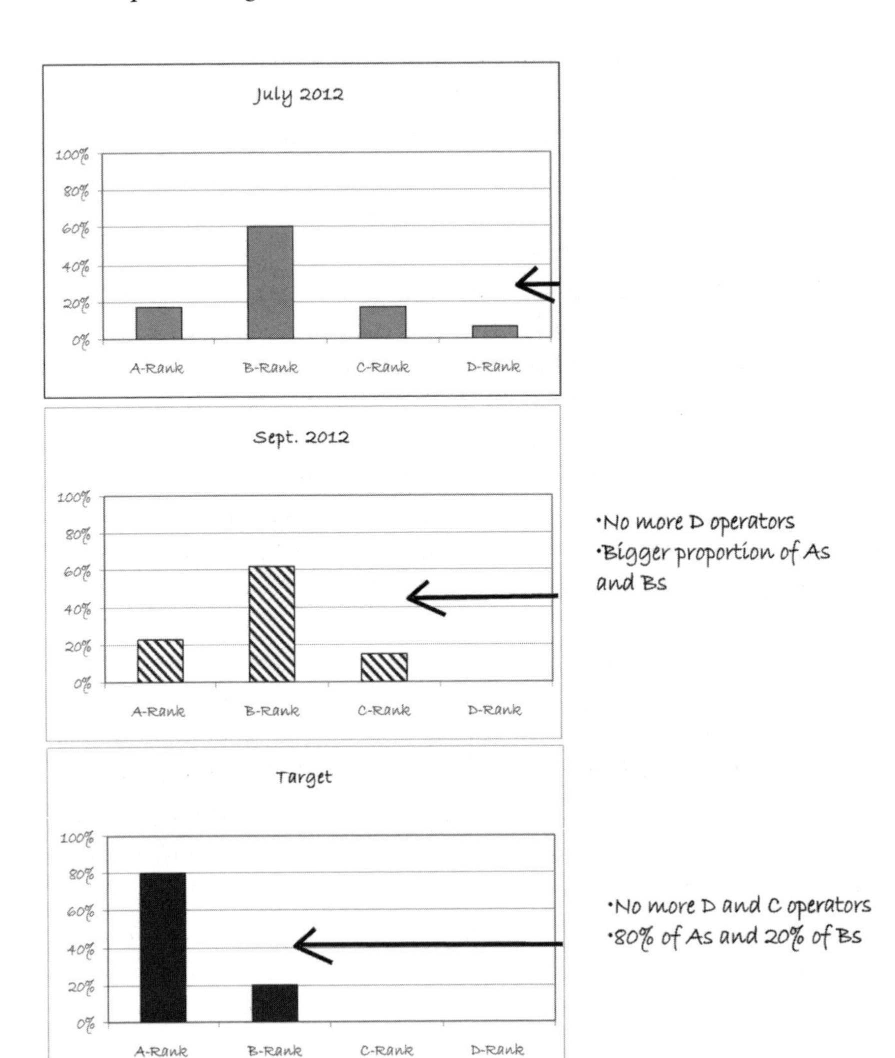

FIGURE 11.4
Improvement and target for operators' performance.

Thomas was really proud of the achievement his associates had made. Daniel was satisfied as well that the plant was starting to be profitable. He then told Thomas the road would be long and paved with obstacles, but concluded, "As Vince Lombardi once said: 'Perfection is not attainable. But, if we chase perfection, we can catch excellence.'"

<u>The 10 steps of OPM</u>

1. In the selected area, list all operators, all machines and all variants of products
2. Identify all possible combinations of an operator working on a specific machine to produce a variant of product. This is the qualification matrix.
3. Measure the "top parts" for each combination recorded in the qualification matrix
4. Use the previous data to draw a histograms with a bar range of 5 seconds for each combination in the qualification matrix
5. Find the mode bar for each combination and record them in the qualification matrix.
6. Compute the variability index for each combination in the qualification matrix.
7. Draw a histogram chart with the variability index on the x-axis: minimum 0% and maximum 100% and a range of 5%. Superimposed all the configurations with the same variability index in the same column (y-axis).
8. Calculate the proportion of operators ranked A, B, C and D.
9. Calculate the overall average variability index of the all operators
10. Set-up your target and move to the next step of Standardized Work to improve...

FIGURE 11.5
Summary of the 10 steps of Operator Performance Mapping (OMP).

Index

About the Author

Alain Patchong is the Director of Assembly at Faurecia Automotive Seating, France. He also holds the title of Master Expert in Assembly processes. He was previously the Industrial Engineering Manager for Europe, the Middle East, and Africa at Goodyear in Luxembourg. In this position, he developed training materials and led a successful initiative for the deployment of Standardized Work in several Goodyear plants.

Before joining Goodyear, he worked with PSA Peugeot Citroën for 12 years where he developed and implemented methods for manufacturing systems engineering and production line improvement. He also led Lean implementations within PSA weld factories.

He teaches at Ecole Centrale Paris and Ecole Supérieure d'Electricité, two French engineering schools. He was a finalist of the INFORMS'* Edelman Competition in 2002 and a Visiting Scholar at MIT† in 2004. He is the author of several articles published in renowned journals. His work has been used in engineering and business school courses around the world.

* Institute for Operations Research and Management Sciences.

† Massachusetts Institute of Technology